Zinc Oxide Based Nano Materials and Devices

Edited by Ahmed M. Nahhas

Published in London, United Kingdom

IntechOpen

Supporting open minds since 2005

Zinc Oxide Based Nano Materials and Devices
http://dx.doi.org/10.5772/intechopen.78819
Edited by Ahmed M. Nahhas

Contributors
Nesrin Horzum, Mohamed Elhousseini Hilal, Tuğba Isık, Trilok Kumar Pathak, Mrumun David Tyona, Raymond Taziwa, Luyolo Ntozakhe, Jun Ki Kim, Sanghwa Lee, Germán Pérez-Zúñiga, Gabriel Herrera-Pérez, Rafael Vargas-Bernal, Ana Maria Valenzuela-Muñiz, Ysmael Verde-Gómez, Prof. Dr. Ahmed M Nahhas

First published in London, United Kingdom, 2019 by IntechOpen
IntechOpen is the global imprint of INTECHOPEN LIMITED, registered in England and Wales, registration number: 11086078, The Shard, 25th floor, 32 London Bridge Street
London, SE19SG - United Kingdom
Printed in Croatia

British Library Cataloguing-in-Publication Data
A catalogue record for this book is available from the British Library

Additional hard and PDF copies can be obtained from orders@intechopen.com

Zinc Oxide Based Nano Materials and Devices
Edited by Ahmed M. Nahhas
p. cm.
Print ISBN 978-1-78923-957-7
Online ISBN 978-1-78923-958-4
eBook (PDF) ISBN 978-1-78984-317-0

For EU product safety concerns:
IN TECH d.o.o., Prolaz Marije Krucifikse Kozulić 3, 51000 Rijeka, Croatia, info@intechopen.com or visit our website at intechopen.com.

Meet the editor

Ahmed M. Nahhas received his PhD degree in Electrical Engineering (Electronics) from the University of Pittsburgh, PA, USA, in 2001. He has served in many prestigious leadership positions, including as the dean of the College of Engineering, the head of the Department, and the vice dean. He is currently with the College of Engineering and Islamic Architecture, Umm Al-Qura University. He has participated in reviewing several academic articles and dissertations in the area of electrical and electronic communications. His current research interests include developing new photonic and electronic devices at nanoscales involving various functional materials, such as wide-bandgap semiconductors, nanostructured materials, and fabrication of ZnO and GaN optical devices. Prof. Nahhas received his first patent in 2016 from the United States and is an official reviewer of several journals.

Contents

Preface

This book presents a review of recent advances in ZnO-based nanomaterials and devices. ZnO as a nanomaterial has gained substantial interest in the research area of wide bandgap semiconductors and is considered to be one of the major candidates for electronic and photonic applications. ZnO has distinguished and interesting electrical and optical properties and is considered as a potential material in optoelectronic applications such as solar cells, surface acoustic wave devices, and UV emitters. ZnO's unique properties have attracted several researchers to study its electrical and optical properties. As a nanostructured material, ZnO exhibits many advantages for nanodevices. ZnO nanostructured material has the ability to absorb UV radiation. Recently, ZnO nanostructured devices have gained much attention due to their various potential applications. In this book, Chapter 1 is an introductory chapter that describes recent advances in ZnO nanomaterials and devices. Chapter 2 addresses the synthesis and characterization of ZnO nanostructures. Chapter 3 reports on the green synthesis of ZnO nanostructures. Chapter 4 introduces doped ZnO nanostructures and photovoltaic solar cell applications. Chapter 5 presents the pyrolysis of carbon-doped ZnO nanoparticles for solar cell applications. Chapter 6 discusses surface-enhanced Raman spectroscopy based on ZnO nanorods for biological applications. Chapter 7 presents anodic ZnO-graphene composite materials in lithium batteries.

Dr. Ahmed M. Nahhas
Professor,
Department of Electrical Engineering,
Faculty of Engineering and Islamic Architecture,
Umm Al Qura University,
Makkah, Saudi Arabia

Section 1

Introduction

Chapter 1

Introductory Chapter: Overview of ZnO Based Nano Materials and Devices

Ahmed M. Nahhas

1. Introduction

This chapter presents an introductory review about ZnO nanomaterials and nanodevices. ZnO as a wide band gap semiconductor has received a great attention in many research areas. This is due to the electrical, optical and structural properties of the ZnO. These properties make ZnO as one of the major contenders for many photonic applications. ZnO has distinguished electrical and optical properties. ZnO is considered as a potential contender in optoelectronic applications such as solar cells and ultraviolet (UV) emitters. The nanostructured ZnO material has many applications in the area of nano based devices. The UV light can be absorbed by the ZnO based nano material. This can be used in several optical applications. Currently, the nano structures based on ZnO materials devices have attracted attention due to their wide range applications.

2. ZnO properties

ZnO is classified as a negative (*n*-type) semiconductor material as grown. ZnO is one of group 2–4 semiconductor materials. ZnO has a band gap of 3.37 eV. ZnO also has a high binding energy. The ZnO binding energy is about 60 meV [1]. The ZnO material has a high exciton binding energy and high thermal stability [2]. It also has a high optical gain [2]. These properties made ZnO as one of the most interesting materials for electronic and optoelectronic based devices. On the other hand, the ZnO's high binding energy permits the fabrication of several photonic devices with a high optical efficiency [3]. Also, the short wavelength optoelectronic devices are being made based on ZnO's wide band gap [3].

ZnO is a transparent optical material suitable for the visible wavelength region [4]. ZnO is also one of the potential materials for optoelectronic applications [5–7]. The ZnO's characteristical properties were investigated by many research groups. That leads to the improvement of the electrical and the optical properties of ZnO. Many other properties of ZnO allow variety of applications. These applications include photovoltaics, LEDs, photodetectors, and microelectromechanical systems (MEMs) [8–12].

3. ZnO crystalline structure

ZnO normally has a hexagonal structure. The Zinc atoms are tetrahedrally coordinated to four oxygen atoms [13]. Moreover, there are two crystalline structure of ZnO. These structures are the wurtzite and the Zinc-blende. These two structures

 IntechOpen

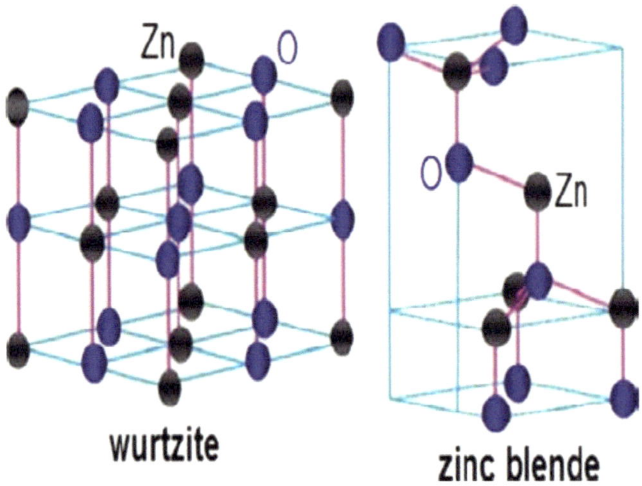

Figure 1.
ZnO different crystalline structures [3].

of ZnO lead to a perfect polar symmetry along the hexagonal axis of the ZnO's crystalline structure. The ZnO based piezoelectricity and spontaneous polarization are due to these crystalline structures. The two ZnO's crystalline structures are shown in **Figure 1**.

4. ZnO doping

Generally, the doping process is used for enhancing the electrical and the optical characteristics of the material [14]. The undoped *n*-type ZnO is usually due to the oxygen vacancies in the ZnO structure [15, 16]. This makes the doping process of ZnO as *p*-type is very difficult. There are several *p*-type dopants for ZnO nano materials. Lithium is one of the *p*-type dopants for the ZnO. It takes a centered position by replacing the zinc atoms in the wurtzite structure of the ZnO [17]. GaN can be used

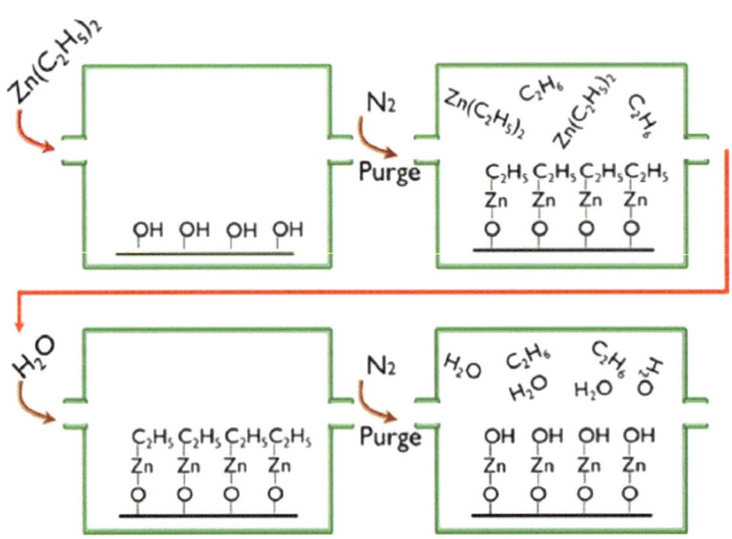

Figure 2.
ZnO doping process [25].

with ZnO to bypass the problem of *p*-type doping of ZnO nano material [18]. GaN is a wide band gap semiconductor with very similar lattice constants to ZnO. The GaN *p*-type dopant is commonly made with magnesium [18]. Several research groups have reported on the fabrication of ZnO hetero-junction devices. These hetero-junctions devices used materials like silicon [19], aluminum gallium nitride [20], and gallium nitride [21] for the *p*-type semiconductors. The electrical resistivity is significantly decreasing for ZnO by using doping elements like aluminum [22], gallium [23], and indium [24]. The ZnO doping process is shown in **Figure 2**.

5. ZnO nanostructured material

Recently, several efforts have been made for developing the ZnO film in different nanoshapes. These nanoshapes include the nanorods, nanowires, and nanocubes [23–26]. These nanoshapes of ZnO films can be used in several applications including the UV laser emission, photodetector and LEDs devices.

The nanoshapes of ZnO nanowires and nanorods based devices are becoming an important part of many potential applications. These applications include the optoelectronics, electromechanical nano devices [27] and biosensing [28].

There are several applications were reported based on ZnO based nanomaterials. These applications include the transparent electronic devices [29], the UV light emitters [30], the piezoelectric devices [31], the p-n junction devices [32], the field effect devices [33], the sensors [34], the optoelectronics [35], and the field emission devices [36].

6. ZnO nanowires based devices

ZnO nanowires based devices have many research interests. Recently, many devices based on ZnO nanostructured nanowires were reported. This is due to the high surface area and the low cost of fabrication [37–40]. There are several fabrication methods being used for the growth of the ZnO nanowires. These methods include the vapor liquid solid [41], metal organic chemical vapor deposition [42], chemical bath deposition [43] and hydrothermal method [44] with varied film qualities. The ZnO nanowires image is shown in **Figure 3**.

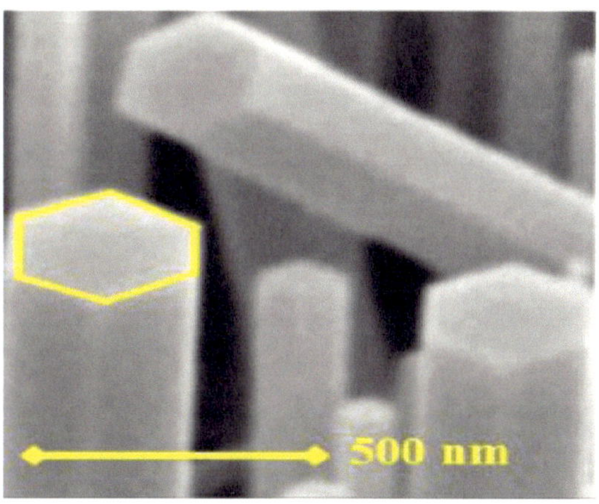

Figure 3.
ZnO nanowires [45].

7. ZnO nanorods based devices

ZnO based nanorods have received a good interest in various optoelectronic nanoscale devices. These devices include the photovoltaic cells, UV laser diodes [46], LEDs, optical sensors, and UV photodetectors [47]. Several applications based on ZnO require a large surface area. The simplest method to increase the surface of the ZnO layer is by growing nanorods. The larger the surface area the more photons are expected to be absorbed by ZnO layer [48]. One way to improve the crystalline structure and the photoluminescence spectra of ZnO nanorods is by annealing process [48].

The ZnO nanorods are being fabricated using several methods. These fabrication methods include the hydrothermal method [49], metal oxide chemical vapor deposition method [50], pulsed laser deposition method [51], aqueous solution method [52]. In these methods, the surface morphologies, optical and electrical properties of ZnO nanorods are depending on the fabrication parameters. These parameters include the deposition time, temperature, and post deposition annealing. The ZnO nanorods are shown in **Figure 4**.

Other fabrication methods of the ZnO nanoshape films are including the sol-gel technique, RF magnetron sputtering, electron beam evaporation, spray pyrolysis (SP), and molecular beam epitaxy, electrochemical deposition [54]. The electrochemical deposition wet method is a well-established solution based process to obtain ZnO nanoshapes thin films [55]. It has many advantages including the low cost and the temperature deposition over other types of fabrication.

Figure 4.
ZnO nanorods [53].

8. ZnO nano based applications

The ZnO material is being used for many optoelectronic based devices applications. These devices applications include the varistors, sensors [13, 56–59], optical wave guides, UV light emitters [60], LEDs [61], micro-electro-mechanical systems [12], spin electronics [62], solar energies [63], p-n junctions [64], field effect and emissions [65–67], displays [68], acoustic waves [69], and solar cells [6].

Author details

Ahmed M. Nahhas
Department of Electrical Engineering, Faculty of Engineering and Islamic Architecture, Umm Al Qura University, Makkah, Saudi Arabia

*Address all correspondence to: amnahhas@uqu.edu.sa

References

[1] Khan I, Khan S, Nongjai R, Ahmed H, Khan W. Hydrothermal synthesis of zinc oxide powders with controllable morphology. Optical Materials. 2013;**35**:1189-1193

[2] Xu H, Wang H, Zhang Y, He W, Zhu M, Wang B, et al. Structural and optical properties of gel-combustion synthesized Zr doped ZnO nanoparticles. Ceramics International. 2004;**30**:93-97

[3] Kung S, Sreenivas K. Defect free C-axis oriented zinc oxide (ZnO) films grown at room temperature using RF magnetron sputtering. AIP Conference Proceedings. 2016;**1731**:1-3

[4] Li Y, Bando Y, Golberg D. ZnO nanoneedles with tip surface perturbations: Excellent field emitters. Applied Physics Letters. 2004;**84**:3603

[5] Saito M, Fujihara S. Large photocurrent generation in dye-sensitized ZnO solar cells. Energy and Environmental Science. 2008;**1**:280-283

[6] Zhou J, Wu X, Xiao D, Zhuo M, Jin H, Luo J, et al. Deposition of aluminum doped ZnO as electrode for transparent ZnO/glass surface acoustic wave devices. Surface and Coating Technology. 2017;**320**:39-46

[7] Choi Y, Kang J, Hwang D, Park S. Recent advances in ZnO based light-emitting diodes. IEEE Transactions on Electron Devices. 2010;**57**:26-41

[8] Zhang M, Gao X, Barra A, Chang P, Huang L, Hellwarth R, et al. Core-shell structured Si/ZnO photovoltaics. Materials Letters. 2015;**140**:59-63

[9] Hossaini H, Moussavi G, Farrokhi M. Oxidation of diazinon in cns-ZnO/LED photocatalytic process: Catalyst preparation, photocatalytic examination, and toxicity bioassay of oxidation by products. Separation and Purification Technology. 2017;**174**:320-330

[10] Sabah M, Hassan M, Naser M, Al-hardan H, Bououdina M. Fabrication of low cost UV photo detector using ZnO nanorods grown onto nylon substrate. Journal of Materials Science. 2015;**26**:1322-1331

[11] Pandya H, Chandra S, Vyas A. Integration of ZnO nanostructures with MEMS for ethanol sensor. Sensors and Actuators B: Chemical. 2012;**161**:923-928

[12] Taube A, Sochacki M, Kwietniewski N, Werbowy A, Gierałtowska S, Wachnicki L, et al. Electrical properties of isotype and anisotype ZnO/4H-SiC heterojunction diodes. Applied Physics Letters. 2017;**110**:1120-1124

[13] Li L, Zhang Y, Yan L, Jiang J, Han X, Deng G, et al. n-ZnO/p-GaN heterojunction light-emitting diodes featuring a buried polarization-induced tunneling junction. AIP Adances. 2016;**6**:125204

[14] Tvarozek V, Shtereva K, Novotny I, Kovac J, Sutta P, Srnanek R, et al. RF diode reactive sputtering of n- and p-type zinc oxide thin films. Vacuum. 2007;**82**:166-169

[15] Liu G, Rahman E, Ban D. Performance optimization of p-n homojunction nanowire based piezoelectric nanogenerators through control of doping concentration. Journal of Applied Physiology. 2017;**118**:94307

[16] Pemmaraju C, Archer T, Hanafin R, Sanvito S. Investigation of n-type donor defects in Co-doped ZnO. Journal of Magnetism and Magnetic Materials. 2007;**316**:e185-e187

[17] Hamid H, Celik-Butler Z. Li-ZnO nanowire carpet as a micro-newton force sensor with nanometer res. In: IEEE Conference on Sensors; 2017. pp. 1-3

[18] Saroj R, Dhar S. Relationship between dislocation and the visible luminescence band observed in ZnO epitaxial layers grown on c-plane p-GaN templates by chemical vapor deposition technique. Journal of Applied Physiology. 2016;**120**:75701

[19] Urgessa N, Dobson S, Talla K, Murape D, Venter A, Botha J. Optical and electrical characteristics of ZnO/Si heterojunction. Physica B: Condensed Matter. 2014;**439**:149-152

[20] Alivov R, Kalinina E, Cherenkov A, Look D, Ataev B, Omaev A, et al. Fabrication and characterization of n-ZnO/p-AlGaN n-ZnO/p-AlGaN heterojunction light-emitting diodes on 6H-SiC substrates. Applied Physics Letters. 2003;**83**:4719

[21] Alvi N, Riaz M, Tzamalis G, Nur O, Willander M. Fabrication and characterization of high-brightness light emitting diodes based on n-ZnO nanorods grown by a low-temperature chemical method on p-4H-SiC and p-GaN. Semiconductor Science and Technology. 2010;**25**:065004

[22] Li Y, Meng J. Al-doping effects on structure and optical properties of ZnO nanostructures. Materials Letters. 2014;**117**:260-262

[23] Chen Y, Huang I, Chang S, Hsueh T. Photodetector of ZnO nanowires based on through-silicon via approach. In: IEEE International Interconnect Technology Conference/Advanced Metallization Conference (IITC/AMC); 2016. pp. 123-124

[24] Yi G, Wang C, Park W. ZnO nanorods: Synthesis, characterization and applications. Semiconductor Science and Technology. 2005;**20**:S22-S34

[25] Viet N, Joao R, Carmen J, Jean-Luc D, Perrine C, Delfina M, et al. Deposition of ZnO based thin films by atmospheric pressure spatial atomic layer deposition for application in solar cells. Journal of Renewable and Sustainable Energy. 2017;**9**:021203

[26] Pan Z, Dai Z, Wang Z. Nanobelts of semiconducting oxides. Science. 2001;**291**:1947-1949

[27] Jianming J, Xiaoqin F, Guibin C. Electromechanical properties of a zigzag ZnO nanotube under local torsion. Journal of Nano Research. 2013;**15**:1-9

[28] Mustafa M, Iqbal Y, Majeed U, Sahdan M. Effect of precursor's concentration on structure and morphology of ZnO nanorods synthesized through hydrothermal method on gold surface. AIP Conference Proceedings. 2017;**1788**:30120

[29] Logothetidis S, Laskarakis A, Kassavetis S, Lousinian S, Gravalidis C, Kiriakidis G. Optical structural properties of ZnO for transparent electronics. Thin Solid Films. 2008;**516**:1345-1349

[30] Pal A, Mohan D. Multi-angle ZnO microstructures grown on Ag nanorods array for plasmon-enhanced near-UV-blue light emitter. Nanotechnology. 2017;**28**:415707-415707

[31] Serhane R, Messaci S, Lafane S, Khales H, Aouimeur W, Bey A, et al. Pulsed laser deposition of piezoelectric ZnO thin films for bulk acoustic wave devices. Applied Surface Science. 2014;**288**:572-578

[32] Nie Y, Deng P, Zhao Y, Wang P, Xing L, Zhang Y, et al. The conversion of PN-junction influencing the

piezoelectric output of a CuO/ZnO nanoarray nanogenerator and its application as a room-temperature self-powered active H_2S sensor. Nanotechnology. 2014;**25**:265501

[33] Tan Q, Wang J, Zhong X, Zhou Y, Wang Q, Zhang Y, et al. Impact of ZnO polarization on the characteristics of metal-ferroelectric-ZnO field effect transistor. IEEE Transactions on Electron Devices. 2011;**58**:2738-2742

[34] Faia P, Furtado C. Effect of composition on electrical response to humidity of TiO2:ZnO sensors investigated by impedance spectroscopy. Sensors and Actuators B: Chemical. 2013;**181**:720-729

[35] Panda D, Tseng T. One-dimensional ZnO nanostructures: Fabrication, optoelectronic properties, and device applications. Journal of Materials Science. 2013;**48**:6849-6877

[36] Zhao Q, Huang C, Zhu R, Xu J, Chen L, Yu D. 2D planar field emission devices based on individual ZnO nanowires. Solid State Communications. 2011;**151**:1650-1653

[37] Lokman A, Arof H, Harun W, Harith Z, Rafaie H, Nor R. Optical fiber relative humidity sensor based on inline Mach-Zehnder interferometer with ZnO nanowires coating. IEEE Sensors Journal. 2016;**16**:312-316

[38] Willander M, Klason P. ZnO nanowires: Chemical growth, electrodeposition, and application to intracellular nano-sensors. Physica Status Solidi C. 2008;**5**:3076-3083

[39] Lupan O, Emelchenko G, Ursaki V, Chai G, Redkin A, Gruzintsev A, et al. Synthesis and characterization of ZnO nanowires for nanosensor applications. Materials Research Bulletin. 2010;**45**:1026-1032

[40] Ramgir N, Kaur M, Sharma P, Datta N, Kailasaganapathi S, Bhattacharya S, et al. Ethanol sensing properties of pure and Au modified ZnO nanowires. Sensors and Actuators B: Chemical. 2013;**187**:313-318

[41] Zhao Q, Klason P, Willander M. Growth of ZnO nanostructures by vapor liquid solid method. Applied Physics A. 2007;**88**:27-30

[42] Pan M, Fenwick W, Strassburg M, Li N, Kang H, Kane M, et al. Metal organic chemical vapor deposition of ZnO. Journal of Crystal Growth. 2006;**287**:688-693

[43] Chiu S, Huang J. Chemical bath deposition of ZnO and Ni doped ZnO nanorod. Journal of Non-Crystalline Solids. 2012;**358**:2453-2457

[44] Polsongkram D, Chamninok P, Pukird S, Chow L, Lupan O, Chai G, et al. Effect of synthesis conditions on the growth of ZnO nanorods via hydrothermal method. Physica B: Condensed Matter. 2008;**403**:3713-3717

[45] Jabr S, Souissi H, Lusson A, Sallet V, Meftah A, Galtier P, et al. The ratio oxygen/zinc effect on photoluminescence emission line at 3.31 eV in ZnO nanowires. Journal of Applied Physics. 2016;**119**:205710

[46] Long H, Fang G, Li S, Mo X, Wang H, Huang H, et al. A ZnO/ZnMgO multiple quantum well ultraviolet random laser diode. IEEE Electron Device Letters. 2011;**32**:54-56

[47] Hwang J, Wang F, Kung C, Chan M. Using the surface plasmon resonance of Au nanoparticles to enhance ultraviolet response of ZnO nanorods based Schottky barrier photodetectors. IEEE Transactions on Nanotechnology. 2015;**14**:318-321

[48] Sipr O, Rocca F. Zn K edge and O K edge x-ray absorption spectra of ZnO surfaces: Implications for nanorods. Journal of Physics: Condensed Matter. 2011;**23**:315501

[49] Lestari A, Iwan S, Djuhana D, Imawan C, Harmoko A, Fauzia V. Effect of precursor concentration on the structural and optical properties of ZnO nanorods prepared by hydrothermal method. AIP Conf. Proc. 2016;**1729**:20027

[50] Montenegro D, Souissi A, Tomas C, Sanjose V, Sallet V. Morphology transitions in ZnO nanorods grown by MOCVD. Journal of Crystal Growth. 2012;**359**:122-128

[51] Mendelsberg R, Kerler M, Durbin S, Reeves R. Photoluminescence behavior of ZnO nanorods produced by eclipse PLD from a Zn metal target. Superlattices and Microstructures. 2008;**43**:594-599

[52] Baliga B. Gallium nitride devices for power electronic applications. Semiconductor Science and Technology. 2013;**28**:74011

[53] Nivedita Y, Veit W. Controlled growth of ZnO nanorods via self-assembled monolayer. Journal of Applied Electrochemistry. 2018; **48**:85-94

[54] Pearton S, Ren F. GaN electronics. Advanced Materials. 2000;**12**:1571-1580

[55] Fu N, Li E, Cui Z, Ma D, Wang W, Zhang Y, et al. The electronic properties of phosphorus doped GaN nanowires from first principle calculations. Journal of Alloys and Compounds. 2014;**596**:92-97

[56] Yuliah Y, Bahtiar A, Fitrilawati SR. The optical band gap investigation of PVP-capped ZnO nanoparticles synthesized by sol-gel method. AIP Conf. Proc. 2016;**1712**:50018

[57] Roh J, Kim H, Park M, Kwak J, Lee C. Improved electron injection in all solution processed n-type organic field effect transistors with an inkjet printed ZnO electron injection layer. Applied Surface Science. 2017;**420**:100-104

[58] Zhang M, Zhang H, Li L, Tuokedaerhan K, Jia Z. Er-enhanced humidity sensing performance in black ZnO based sensor. Journal of Alloys and Compounds. 2018;**744**:364-369

[59] Ling Z, Wen Z. Room-temperature gas sensing of ZnO based gas sensor: A review. Sensors and Actuators, A: Physical. 2017;**267**:242-261

[60] Konenkamp R, Word R, Schegel R. Vertical nanowire light-emitting diode. Applied Physics Letters. 2004;**85**:6004-6006

[61] Mckinstry S, Marult P. Thin film piezoelectrics for MEMS. Journal of ElectroceramicsJournal of Electroceramics. 2004;**12**:7-17

[62] Ushio Y, Miyayama M, Yanagida H. Effects of interface states on gas sensing properties of a CuO/ZnO thin film heterojunction. Sensors and Actuators B: Chemical. 1994;**17**:221-226

[63] Harima H. Raman studies on spintronics materials based on wide bandgap semiconductors. Journal of Physics. Condensed Matter. 2004;**16**:S5653-S5660

[64] Xiang J, Zhu P, Masuda Y, Okuya M, Kaneko S, Koumoto K. Flexible solar cell from zinc oxide nanocrystalline sheets self assembled by an In-situ electrodeposition process. Journal of Nanoscience and Nanotechnology. 2006;**6**:1797-1801

[65] Khachar U, Solanki P, Choudhary R, Phase D, Ganesan V, Kuberkar D. Room temperature positive magnetoresistance and field effect studies of manganite-based heterostructure. Applied Physics A. 2012;**108**:733-738

[66] Boccuzzi F, Chiorino A, Tsubota S, Haruta M. An IR study of CO-sensing

mechanism on Au/ZnO. Sensors and Actuators B: Chemical. 1995;**25**:540-543

[67] Singh J, Patil S, More M, Joag D, Tiwari R, Srivastava O. Formation of aligned ZnO nanorods on self grown ZnO template and its enhanced field emission characteristics. Applied Surface Science. 2010;**256**:6157-6153

[68] Alvi M, Al-Ghamdi A, Husain M. Field emission studies of CNTs/ZnO nanostructured thin films for display devices. Physica B: Condensed Matter. 2017;**521**:312-316

[69] Lao C, Gao P, Yang R, Zhang Y, Dai Y, Wang Z. Formation of double side teethed nanocombs of ZnO and self catalysis of Zn terminated polar surface. Chemical Physics Letters. 2006;**417**:358-362

Section 2

ZnO Nanostructures

Chapter 2

Structural and Luminescence Properties of ZnO Nanoparticles Synthesized by Mixture of Fuel Approach in Solution Combustion Method

Trilok K. Pathak and H.C. Swart

Abstract

Zinc oxide has been used for many applications, for example optoelectronic devices, ceramics, catalysts, pigments, varistors and many other important applications. In this study, ZnO nanoparticles were synthesized by mixture of fuel approach in solution combustion method. Mixtures of urea, glycine and citric acid were mixed at room temperature with Zinc nitrates as fuels resulting in spontaneous ignition resulting in production of ZnO nanopowder. The crystal structure and size of the synthesized powder were determined by X-ray diffractometer (XRD), which revealed that the synthesized ZnO nanopowder has the pure wurtzite structure having crystallite size 26–40 nm. Optical studies of nanomaterial were examined by FTIR and UV-Visible absorption spectrum. The luminescence studies also investigated in the visible region 360–800 nm with excitation 325 nm laser. These nanomaterials may be used in solid-state lightening devices.

Keywords: ZnO NPs, X-ray diffraction, bandgap, luminescence

1. Introduction

Nanoscale ZnO powder has attracted great attention due to its excellent physical and chemical properties they are widely used in nanoscale devices such as nanogenerators [1], ultraviolet photodetectors [2], gas sensors [3], solar cells [4], field emission displays [5], electrical and optical devices [6, 7], photocatalysis [8, 9], medical [10] and environmental applications [11]. These nanomaterials have novel electronic, structural and thermal properties which have potential interest in basic and applied research. ZnO is a wide bandgap (E_g = 3.37 eV) semiconductor some basic properties listed in **Table 1** [12].

Semiconductor nanocrystals or nanoparticles may have superior optical and antibacterial properties than bulk crystals due to quantum confinement effects and the large surface to volume ratio. The synthesis and properties of ZnO nanostructure such as nanowires [13], nanotubes [14], nanorods [15] and nanoparticles [16] have been reported. The nanoparticles have great significance

Property	Measured value
Crystal structure	Hexagonal, wurtzite
Molecular weight	Zn:65.38, O:16 and ZnO:81.38
Lattice constant	a = 3.246 Å, c = 5.207 Å
Density	5.67 g/cm^3 or 4.21 × 10^{19} ZnO molecules/mm^3
Cohesive energy	E_{coh} = 1.89 eV
Melting point	T_m = 2250 K
Heat of fusion	4470 cal/mole
Thermal conductivity	25 W/mK at 20°C
Bandgap at RT	3.37 eV
Refractive index	2.008
Electron and hole effective mass	m_e^* = 0.28, m_h^* = 0.59
Dielectric constant	ε_o = 8.75, ε_∞ = 3.75
Exciton binding energy	E_b = 60 meV

Table 1.
Basic properties of ZnO [12].

as three dimensional confined systems bridging the gap between bulk materials and molecular compounds. A variety of techniques have been employed for the synthesis of ZnO nanoparticles such as sol-gel synthesis [17], the hydrothermal method [18], the solution combustion method [19] and solid state reactions [20]. Among these, the combustion technique is noteworthy as a fast method to synthesize nanocrystalline materials in as-synthesized form with large surface area without the further need of heat treatment. Nanocrystalline oxides are produced through the redox reaction between an oxidizer containing the metal precursor and anorganic fuel at a moderately low initiation temperature of around 350–600°C within a few minutes [21]. The main advantage of this method is that the high temperature of the exothermic reaction assures high purity and well crystallized powder. In combustion synthesis, the type of fuel and the fuel to oxidizer ratio (F/O) play critical roles in influencing the nature of combustion reaction and the flame temperature. Selection of a suitable fuel and the F/O ratio influences the combustion process and the properties of the product. The F/O ratio of unity is known to produce highest exothermicity with complete combustion. An arbitrary ratio of fuel to oxidizer (F/O—1) sometimes leads to formation of intermediate phases raw materials in the final product [22]. In this regard, various fuels have been tested to synthesize nanocrystalline ZnO. Sousa et al. [23] used metallic nitrate and urea to synthesis ZnO nanopowder with a size about 400–500 nm for various applications. Hwang et al. [24] worked on ZnO nanopowder synthesized by a combustion method with glycine as a fuel and metal nitrate mixed in a stoichiometric ratio.

In the present work, we report the synthesis of nanocrystalline ZnO powders by combustion technique using new, eco-friendly and cost-effective organic fuels as urea, glycine and citric acid. The effect of fuel in different ratio of two fuels combinations on the properties of the final product has been studied. The structure and luminescence properties of ZnO nanoparticles are also being studied in this work.

2. Experimental

2.1 Preparation of ZnO nanoparticles

Synthesis of ZnO NPs the different materials were used such as zinc nitrate, urea, glycine, and citric acid. **Table 2** shows the characteristic of the raw materials. The chemical reaction used in synthesis of ZnO powder are given in **Table 3**.

The zinc nitrate hexahydrate and fuel were dissolved in 5 ml of double distilled water and stirred thoroughly to obtain a transparent solution, which was placed inside a preheated muffle furnace at 600°C to initiate the combustion process. Within a short time the mixture ignited with a flame and the rapid evolution of enormous amounts of gases produced a voluminous foamy product (ash). This was ground using an agate pestle and mortar to produce the final powder, without any additional heat treatment. The fuels used in this synthesis have different combination of fuels and shown in **Table 4**.

The synthesis process of ZnO NPs is illustrated in **Figure 1**.

Raw materials	Formulation	Molecular weight (g/mol)	Manufacturer
Zinc nitrate	$Zn(NO_3)_2.6H_2O$	297.49	Sigma Aldrich
Urea	NH_2CONH_2	60.06	Sigma Aldrich
Glycine	$NH_2.CH_2.COOH$	75.06	Sigma Aldrich
Citric acid	$C_6H_8O_7.H_2O$	210.14	Sigma Aldrich

Table 2.
Characteristics of raw material.

Fuel	Combustion reaction
Urea	$3Zn(NO_3)_2 + 5CO(NH_2)_2 \rightarrow 3ZnO + 5CO_2 + 10H_2O + 8N_2$
Glycine	$Zn(NO_3)_2 + 2CH_2(NH_2)(COOH) + 2O_2 \rightarrow ZnO + 4CO_2 + 5H_2O + 2N_2$
Citric acid	$Zn(NO_3)_2 + C_3H_5O(COOH)_3 + 2O_2 \rightarrow ZnO + 6CO_2 + 4H_2O + N_2$

Table 3.
Chemical reaction in combustion synthesis of ZnO using different fuels [25].

Sample name	Fuels contents (%)		
	Urea	Glycine	Citric acid
UC1	75	—	25
UC2	50	—	50
UC3	25	—	75
UG1	75	25	—
UG2	50	50	—
UG3	25	75	—

Table 4.
Sample name with respect to used fuels in different combination of fuels.

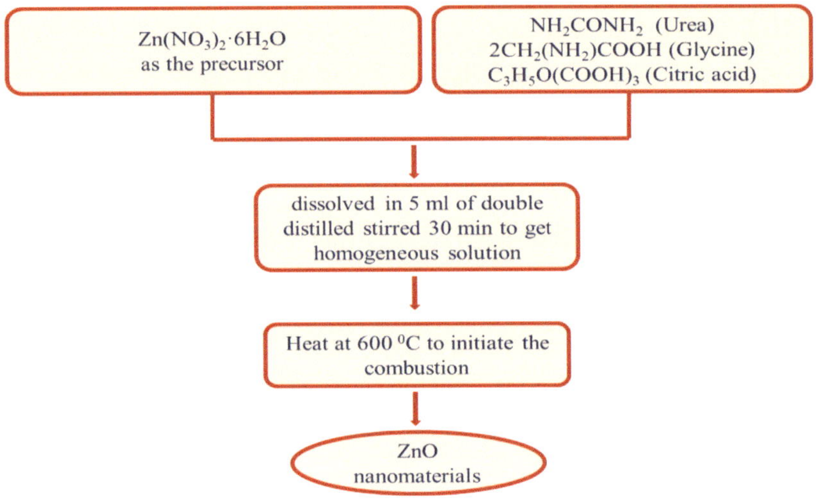

Figure 1.
Systematic diagram of ZnO synthesized by the combustion method.

2.2 Characterization method

The prepared ZnO-NPs were characterized by X-ray diffraction (XRD) using advanced Bruker D8 diffractometer with Cu K_α radiation was carried out to check up the crystal structure. The bond characteristics studies using FTIR-8400S. The Optical transmittance spectra were collected using a UV-Vis-IR spectrophotometer (Perkin Elmer, lambda 950). The photoluminescence (PL) data was recorded using 325 nm He-Cd laser system.

3. Results and discussion

3.1 X-ray diffraction pattern

The XRD patterns of the ZnO powders synthesized using mixed fuels is depicted in **Figure 2a** and **b** and are typical of ZnO powders having the hexagonal wurtzite structure (JCPDS 01-036-1451). This indicates that the ZnO was formed directly by the self-propagating high temperature exothermic combustion reaction initiated at moderate temperature. The crystallize size varied from 30 to 70 nm with different fuels contents. UC2 and UG3 show the wurtzite ZnO structure without any impurity peak in the XRD pattern. All three fuels resulted in nanocrystalline powders, but the crystallite size varied significantly with the type of fuel. The effect of the type of fuel, and the F/O ratio in the case of urea, on the properties of the final product also has been studied in our previous research article [25].

3.2 Fourier transform infrared spectroscopy (FTIR)

The FTIR spectrum of ZnO is shown in **Figure 3**. The broad band with very low intensity at 3466 cm^{-1} corresponding to the vibration mode of water OH group indicating the presence of small amount of water adsorbed on the ZnO nanocrystal surface during synthesis. A strong band at 482–455 cm^{-1} is attributed to the Zn-O stretching band. The bond related to C=O and other are shown in **Table 5**.

3.3 UV-Visible absorption spectrum

The UV-Vis reflectance spectra of the ZnO nanomaterial synthesized using different fuels are shown in the inset of **Figure 4a** and **b**, and the corresponding absorbance spectra are calculated using the Kubelka-Munk function [26]:

$$K = \frac{(1 - R)^2}{2R} \quad (1)$$

where K is the reflectance transformed according to Kubelka-Munk, h is Planck constant, υ is the light frequency, and R is reflectance (%). The relevant increase in the absorption at wavelengths less than 400 nm can be assigned to the optical band-gap absorption of ZnO due to changes in their morphologies, particle size and surface microstructures or the quantum confinement effect [27, 28]. The absorption edges are change with fuels as taken to synthesis ZnO. ZnO has a direct transition

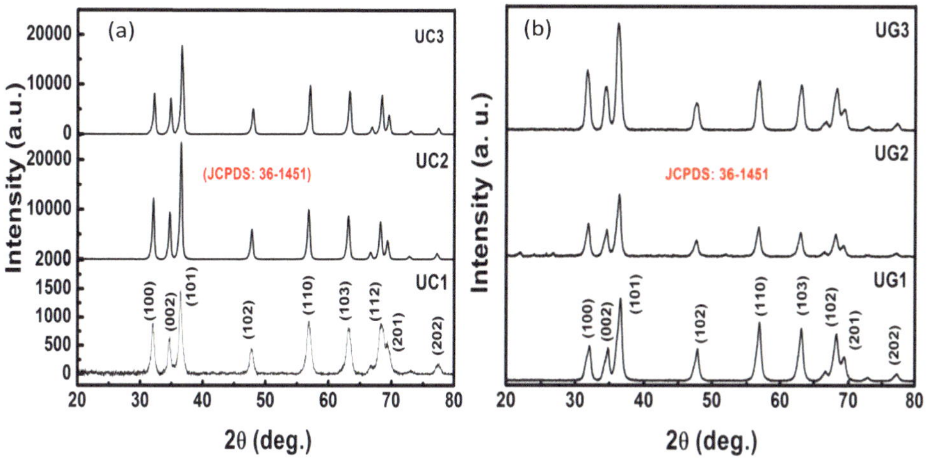

Figure 2.
XRD pattern: (a) ZnO with mixed fuel (urea + citric acid) and (b) ZnO with mixed fuel (urea + glycine).

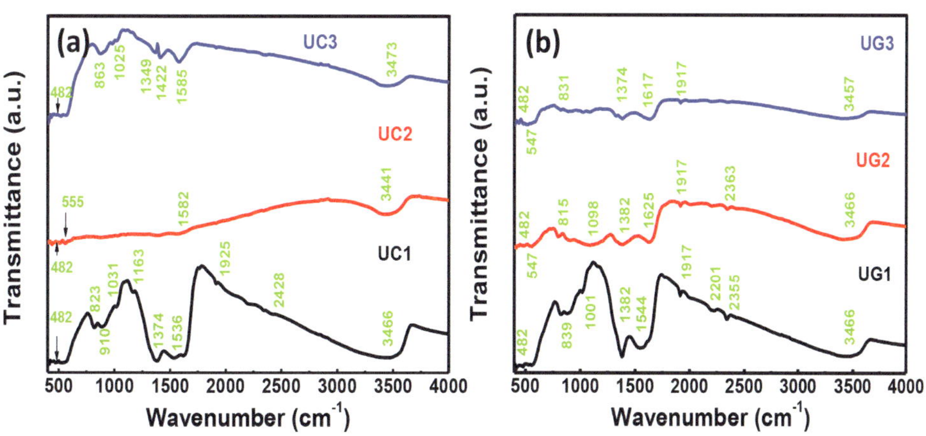

Figure 3.
FTIR spectrum: (a) ZnO with mixed fuel (urea + citric acid) and (b) ZnO with mixed fuel (urea + glycine).

Functional group	Wavenumber (cm^{-1})					
	ZC1	ZC2	ZC3	ZG1	ZG2	ZG3
Stretching vibration of Zn-O	482	482–555	482	482	482–547	482–547
Bending mode of carbonate	823–910	—	863	839	815	831
Stretching vibration of C-O	1031	—	1025	1001	1098	—
C-H in plane bending vibration	1163	—	—	—	—	—
Bending vibration of $-CH_2$	1374	—	1349	1382	1382	1374
C=O band	1536	1582	1585	1544	1625	1617
Carboxyl group	1925	—	—	1917	1917	1917
Existence of CO_2	2428	—	—	2201–2355	2363	—
O-H stretching of water	3466	3441	3473	3466	3466	3457

Table 5.
Chemical boding characteristics of synthesized ZnO with mixed fuels.

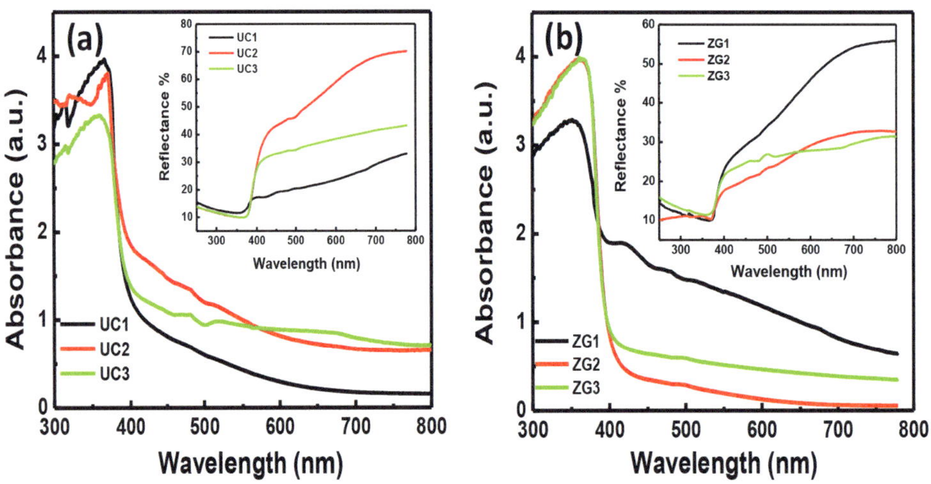

Figure 4.
Absorbance spectra with reflectance insect: (a) ZnO with mixed fuel (urea + citric acid) and (b) ZnO with mixed fuel (urea + glycine).

and the corresponding bandgaps for different mixed fuels shown in **Figure 5a** and **b** respectively are calculated from a Tauc plot of $(\alpha h\nu)^2$ versus the photon energy $(h\nu)$. These bandgap values blue shifted little 3.08 to 3.2 eV relative to the zinc oxide nanomaterial.

3.4 Photoluminescence study

The photoluminescence properties of semiconductor materials undergo change when their size gets down to nanometer scale known as the quantum size effects. The photoluminescence originates from the recombination of surface states. **Figure 6a** and **b** shows the photoluminescence spectra of ZnO powder synthesis by different fuels with excitation wavelength of 325 nm at room temperature. The spectra exhibits two emission peaks, One is located at around 384 nm (UV region)

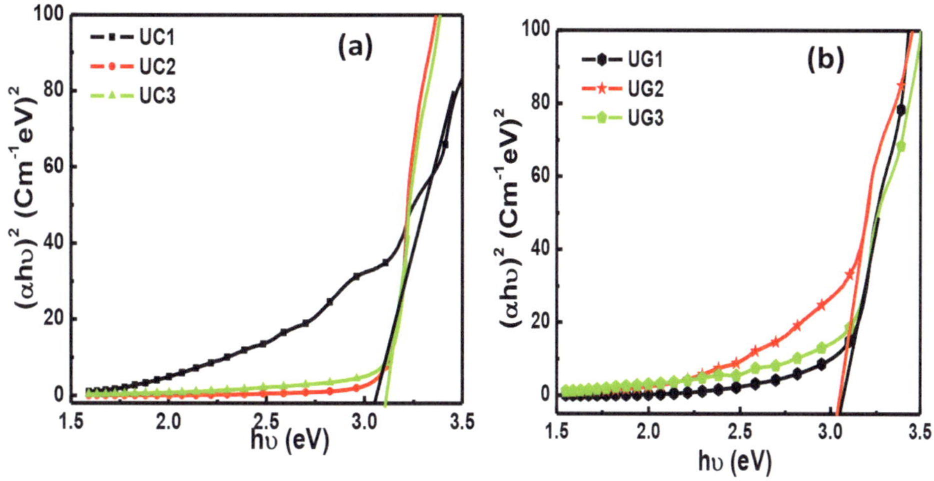

Figure 5.
Energy bandgap: (a) ZnO with mixed fuel (urea + citric acid) and (b) ZnO with mixed fuel (urea + glycine).

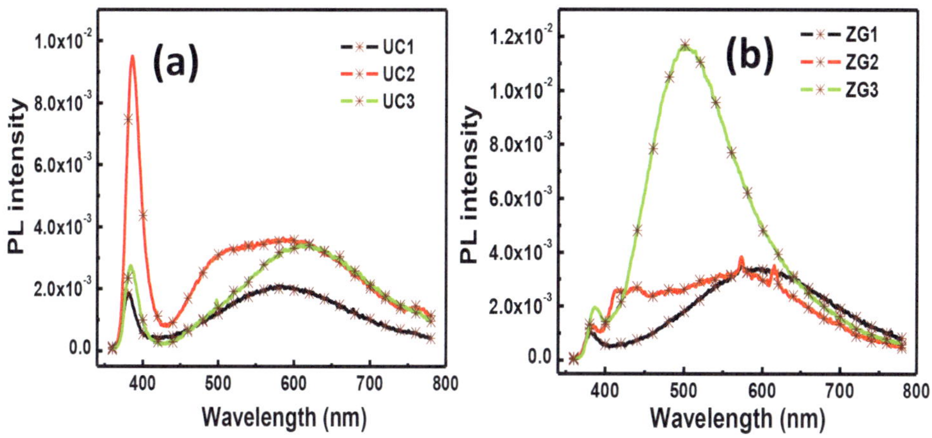

Figure 6.
PL spectra (a) ZnO with mixed fuel (Urea + Citric acid) (b) ZnO with mixed fuel (urea+glycine).

corresponding to the near-band-edge emission [29] which originates from free exciton emission and the other peak corresponding to ionized oxygen vacancies [30] with change for different fuels. High intensity oxygen vacancies peak at 632 nm is obtained for ZnO nanoparticle synthesis by urea with citric acid and band to band peak is eliminate. UC2 shows maximum band edge intensity and ZG3 shows maximum defect related emission.

4. Conclusions

ZnO nanomaterials were successfully synthesized by the combustion method using different fuels ratio. The XRD patterns were consistent with polycrystalline ZnO having the hexagonal wurtzite structure. The ZnO NPs size changed for different fuels with the minimum crystallite size of 26–40 nm obtained by using Glycine, citric acid with Urea at different ratio. The chemical band study shows that OH group has least intensity at higher Urea content. There is little change observed in bandgap with different fuel contents. In the ZnO powder synthesized with different fuels using glycine the band to band PL peak intensity was negligible compared to the defect related emission.

Acknowledgements

This work is based on the research supported by Department of Physics, University of the Free State, Bloemfontein, South Africa. The PL system used in this study is supported both technically and financially by rental poll programme of the national laser centre (NCL). The author states that some part related to this work is published in RSC Journal. The author has permission from the publisher Journal RSC advances to use parts of his previously published work titled "Effect of fuel content on luminescence and antibacterial properties of zinc oxide nanocrystalline powders synthesized by the combustion method."

Author details

Trilok K. Pathak[1,2]* and H.C. Swart[2]

1 Department of Physics, TKCOE Teerthanker Mahaveer University, Moradabad, India

2 Department of Physics, University of Free State, Bloemfontein, South Africa

*Address all correspondence to: tpathak01@gmail.com

References

[1] Wang X, Song J, Wang ZL. Nanowire and nanobelt arrays of zinc oxide from synthesis to properties and novel devices. Journal of Materials Chemistry. 2007;**17**:711-720. DOI: 10.1039/B616963P

[2] Jun JH, Seong H, Cho K, Moon BM, Kim S. Ultraviolet photodetectors based on ZnO nanoparticles. Ceramics International. 2009;**35**:2797-2801. DOI: 10.1016/j.ceramint.2009.03.032

[3] Lin HM, Tzeng SJ, Hsiau PJ, Tsai WL. Electrode effects on gas sensing properties of nanocrystalline zinc oxide. Nanostructured Materials. 1998;**10**:465-477. DOI: 10.1016/s0965-9773(98)00087-7

[4] Wang ZS, Huang CH, Huang YY, Hou YJ, Xie PH, Zhang BW, et al. A highly efficient solar cell made from a dye-modified ZnO-covered TiO_2 nanoporous electrode. Chemistry of Materials. 2001;**13**:678-682. DOI: 10.1021/cm000230c

[5] Kitano M, Shiojiri M. Benard convection ZnO/resin lacquer coating—A new approach to electrostatic dissipative coating. Powder Technology. 1997;**93**:267-273. DOI: 10.1016/S0032-5910(97)03283-X

[6] Zheng MJ, Zhang LD, Li GH, Shen WZ. Fabrication and optical properties of large-scale uniform zinc oxide nanowire arrays by one-step electrochemical deposition technique. Chemical Physics Letters. 2002;**363**:123-128. DOI: 10.1016/S0009-2614(02)01106-5

[7] Wu R, Xie CS. Formation of tetrapod ZnO nanowhiskers and its optical properties. Materials Research Bulletin. 2004;**39**:637-645. DOI: 10.1016/j.materresbull.2003.12.009

[8] Curridal ML, Comparelli R, Cozzli PD, Mascolo G, Agostiano A. Colloidal oxide nanoparticles for the photocatalytic degradation of organic dye. Materials Science and Engineering C. 2003;**23**:285-289. DOI: 10.1016/S0928-4931(02)00250-3

[9] Kamat VP, Huehn R, Nicolaescu R. A "sense and shoot" approach for photocatalytic degradation of organic contaminants in water. The Journal of Physical Chemistry. B. 2002;**106**:788-794. DOI: 10.1021/jp013602t

[10] Rasmussen JW, Martinez E, Louka P, Wingett DG. Zinc oxide nanoparticles for selective destruction of tumor cells and potential for drug delivery applications. Expert Opinion on Drug Delivery. 2010;**7**(9):1063-1077. DOI: 10.1517/17425247.2010.502560

[11] Udom I, Ram MK, Stefanakos EK, Hepp AF, Goswami DY. One dimensional-ZnO nanostructures: Synthesis, properties and environmental applications. Materials Science in Semiconductor Processing. 2013;**16**(6):2070-2083. DOI: 10.1016/j.mssp.2013.06.017

[12] Janotti A, Van de walle CG. Fundamental of zinc oxide as a semiconductor. Reports on Progress in Physics. 2009;**72**:126501-126529. DOI: 10.1088/0034-4885/72/12/126501

[13] Kong YC, Yu DP, Zhang B, Fang W, Feng SQ. Semiconductor nanowires-ideal bridge between macro Newton world and micro quantum world. Applied Physics Letters. 2001;**78**:407. DOI: 10.1063/1.1342050

[14] Xing YJ, Xi ZH, Xue ZQ, Zhang XD, Song JH, Wang RM, et al. Optical properties of the ZnO nanotubes synthesized via vapor phase growth. Applied Physics Letters. 2003;**83**:1689. DOI: 10.1063/1.1605808

[15] Zhang BP, Binh NT, Segawa Y, Kashiwaba Y, Haga K. Photoluminescence study of ZnO nanorodsepitaxially grown on sapphire (1120) substrates. Applied Physics Letters. 2004;**84**:586-588. DOI: 10.1063/1.1738520

[16] Guo L, Yang S, Yang C, Yu P. Highly monodisperse polymer-capped ZnO nanoparticles: Preparation and optical properties. Applied Physics Letters. 2000;**76**:2901. DOI: 10.1063/1.126511

[17] Chu SY, Yan TM, Chen SL. Characteristics of sol–gel synthesis of ZnO based powders. Journal of Materials Science Letters. 2000;**19**:349-352. DOI: 10.1177/1464420714550192

[18] Liu B, Zeng HC. Hydrothermal synthesis of ZnO nanorods in the diameter regime of 50 nm. Journal of the American Chemical Society. 2003;**125**:4430-4431. DOI: 10.1021/ja0299452

[19] Aimable A, Buscaglia MT, Bowen P. Polymer-assisted precipitation of ZnO nanoparticles with narrow particle size distribution. Journal of the European Ceramic Society. 2010;**30**:91-598. DOI: 10.1179/1743280411Y

[20] Sun ZP, Liu L, Zhang L, JiaD Z. Rapid synthesis of ZnO nano-rods by onestep, room-temperature, solid-state reaction and their gas-sensing properties. Nanotechnology. 2006;**17**:2266-2270. DOI: 10.1088/0957-4484/17/9/032

[21] Patil KC, Aruna ST, Mimani T. Combustion synthesis: An update. Solid State & Materials Science. 2002;**6**:507-512. DOI: 10.1016/S1359-0286(02)00123-7

[22] Noori NR, Mamooryb RS, Alizadehb P, Mehdikhani A. Synthesis of ZnO nano powder by a gel combustion method. Journal of Ceramic Processing Research. 2008;**9**(3):246-249. DOI: 10.1016/S1470-1804(99)80057-1

[23] Sousa de VC, Morelli MR, Kiminami RHG. Combustion process in the synthesis of ZnO-Bi$_2$O$_3$. Ceramics International. 2000;**26**:561-564

[24] Hwang CC, Wu TY. Synthesis and characterization of nanocrystalline ZnO powders by a novel combustion synthesis method. Materials Science and Engineering B. 2004;**111**:197-206

[25] Pathak TK, Kumar A, Swart CW, Swart HC, Kroon RE. Effect of fuel content on luminescence and antibacterial properties of zinc oxide nanocrystalline powders synthesized by the combustion method. RSC Advances. 2016;**6**:97770-97782. DOI: 10.1039/C6RA22341A

[26] Yu J, Li C, Liu S. Effect of PSS on morphology and optical properties of ZnO. Journal of Colloid and Interface Science. 2008;**326**:433-438. DOI: 10.1016/j.jcis.2008.07.052

[27] Pathak TK, Swart HC, Kroon RE. Structural and plasmonic properties of noble metal doped ZnO nanomaterials. Physica B. 2018;**535**:114-118. DOI: 10.1016/j.physb.2017.06.074

[28] Dressel M, Gruner G. Electrodynamics of Solids, Optical Properties of Electron in Matter. Cambridge: Cambridge University Press; 2002. pp. 159-165. DOI: 10.1017/CBO9780511606168

[29] Huang MH, Wu Y, Feick H, Tran N, Weber E, Yang P. Catalytic growth of zinc oxide nanowires by vapor transport. Advanced Materials. 2001;**13**(2):113-116. DOI: 10.1002/1521-4095(200101)13:2<113::aid-adma113>3.0.co;2-h

[30] Pathak TK, Kroon RE, Swart HC. Photocatalytic and biological applications of Ag and Au doped ZnO nanomaterial synthesized by combustion. Vacuum. 2018;**157**:508-513. DOI: 10.1016/j.vacuum.2018.09.020

Chapter 3

Green Synthesis of Zinc Oxide Nanostructures

Tuğba Isık, Mohamed Elhousseini Hilal and Nesrin Horzum

Abstract

ZnO-based nanomaterials have been proven to be of great use for several leading applications since the beginning of nanoscience due to the abundance of zinc element and the relatively easy conversion of its oxide to nanostructures. Nowadays, ZnO as nanoparticles, nanowires, nanofibers as well as plenty of other sophisticated nanostructures takes place among the pioneer nanomaterials employed in the photovoltaic systems, fuel cells, and biomedical fields. Nevertheless, optimizing energy consumption and being eco-friendly are the challenging requirements that are still to be overcome for their synthesis. Green chemistry has been strongly presented recently in the scientific arena as an adequate potential alternative; worldwide investigations have been held on subjects involving bacteria, fungus, or algae-based synthesis as efficient options, and some of the intriguing scientific findings on this subject are reported hereafter.

Keywords: biosynthesis, hydrothermal, microwave, nontoxic, sonochemical

1. Introduction

There are many conventional zinc oxide (ZnO) nanostructure synthesis routes employing the chemical and physical methods, which require particular set-up, high cost, high temperature-pressure conditions, and nonecological chemicals [1]. However, high-energy consumption of these routes and released toxic chemicals after the process can be hazardous to the environment and human health. In recent years, the green synthesis approach has been gaining attention, which eliminates the use of toxic chemicals and applies environmentally friendly routes. These strategies handle the use of plant-extracts, microorganisms, biomolecules, and ionic liquids by applying hydrothermal, microwave-assisted, sonochemical, and low-temperature processes (**Figure 1**).

The aim of these developments is to allow the use of toxic chemicals and reduce energy consumption by using simple, rapid, and safe routes. Green synthesis strategies for the ZnO nanostructures could be summarized as biosynthesis (natural extract-based, microorganism-based, and biomolecule-based) and nontoxic chemical synthesis (water-based, calcination, solvent-free, and ionic liquid).

2. Biosynthesis of ZnO nanostructures

2.1 Natural extract-based ZnO nanostructures

Natural extracts (mainly phytochemicals) obtained from plants, leaves, fruit peels, flowers, and seeds have been utilized for the green synthesis of metal oxide nanoparticles for years. After the plants are collected from different sources, they are washed with water and basic extraction procedures are applied to obtain plant extracts in which leaves are ground and immersed in water by stirring at room temperature for a while. Then, the solutions are filtered and the eluted extract solution is separated for further use in ZnO synthesis (**Figure 2**). The eluent solution could be used directly for ZnO synthesis or could be dried for the concentration of solid extracts. Afterward, zinc precursors and plant extracts are reacted under various pH and temperature conditions [2]. If the extract is used as an aqueous solution, the zinc precursors are added into the solution. Otherwise, the zinc precursor and powder form of leaf extract are mixed in distilled water. The key mechanism is the oxidation and reduction of metal ion 'zinc' by phytochemicals, which are found in natural extracts. The leaf extracts behave as reducing and capping agents. Under favor of plant extracts, the synthesis procedure can be accomplished without using any chemical stabilizers. Finally, the obtained powders are washed with methanol or ethanol and annealed at high temperatures to attain purity [3].

The green synthesized ZnO nanoparticles have been used in various fields such as biomedical application due to their significant antibacterial activities, photocatalysis, and metal ion adsorption purposes [4]. Moreover, nanoparticles synthesized by the green route exhibit better antibacterial performances due to the functional

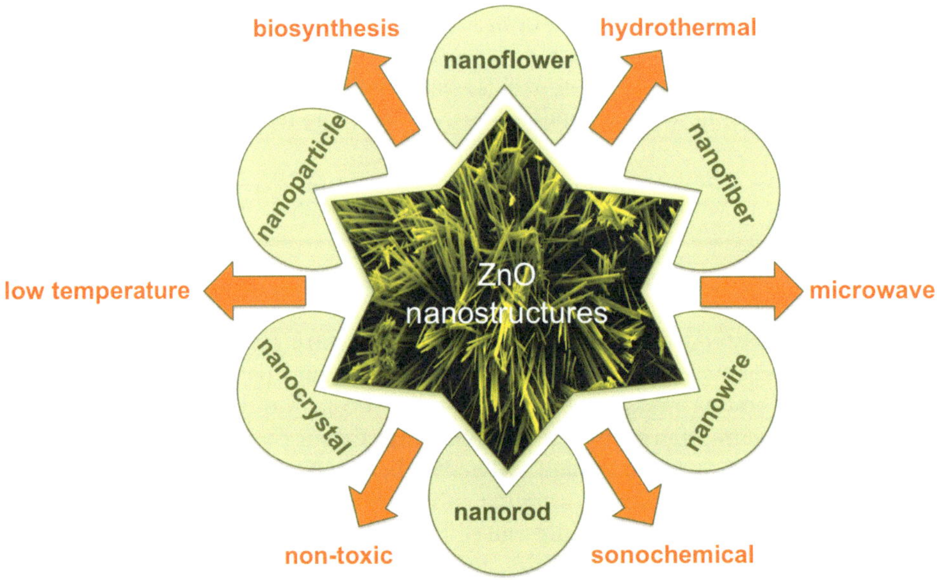

Figure 1.
Green synthesis strategies of ZnO nanostructures with various morphologies.

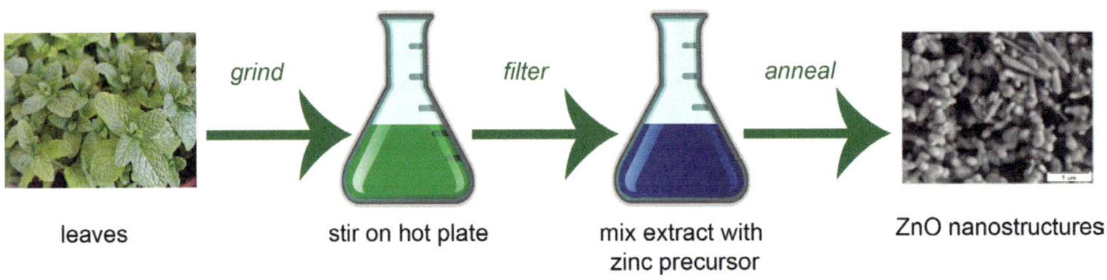

Figure 2.
Synthesis route of ZnO nanostructures from leaf extracts.

groups on their surfaces that come from phytochemicals. Here, we will describe the main applications of natural extract-based green synthesized ZnO nanoparticles.

2.2 Biomedical applications

The advantage of using natural extracts for the synthesis of ZnO nanoparticles is that coating of nanoparticles with various pharmacologically active biomolecules on the metal oxide surface allows the conjugation of nanoparticles with receptors of the bacterial membrane. These molecules might be flavones, aldehydes, amides, polysaccharides, etc. and the green synthesized nanoparticles exhibit better biomedical activity than the chemically synthesized ones [1]. Inorganic metal oxides have widely emerged as antibacterial, antioxidant, antifungal, and anticancer agents in the last decades. Moreover, because of their specific targeting and nominal toxicity, the metal oxide nanoparticles could be used in personalized medicine applications. In the area of nanoscaled metal oxides, ZnO has shown promising activity in the biomedical field due to its unique electronic, optical, and medicinal properties. The ZnO nanoparticles show antibacterial activity against a broad spectrum of pathogenic bacteria, and these nanoparticles adopt various mechanisms such as reactive oxygen species (ROS) generation, cell membrane integrity disruption, biofilm formation, or enzyme inhibition [5]. Under UV irradiation, ROS such as superoxide ions, hydroxyl ions, singlet oxygen species, and peroxide molecules are formed. The formed peroxide ions could easily penetrate through the cell membrane and result in cell death. **Figure 3** shows the possible ROS generation mechanism and its effect on the bacterial cell wall.

Cell membrane integrity disruption is another significant mechanism for the antibacterial effect of ZnO nanoparticles. Penetration of ZnO nanoparticles results in cell death by the loss of phospholipid bilayer integrity and leakage of intracellular components of the cell. While the Gram-positive bacteria have a thick layer of peptidoglycan, teichoic acid, and lipoteichoic acid in their cell membrane, Gram-negative bacteria have a triple layer of peptidoglycan. The different structure of cell membranes of these two types of bacteria results in a different mechanism of nanoparticle penetration through the cell membranes. In this part, we focused on the biomedical activity of ZnO nanoparticles, and in **Table 1**, the used plant extracts, zinc precursors, biomedical applications and related biomolecules are summarized.

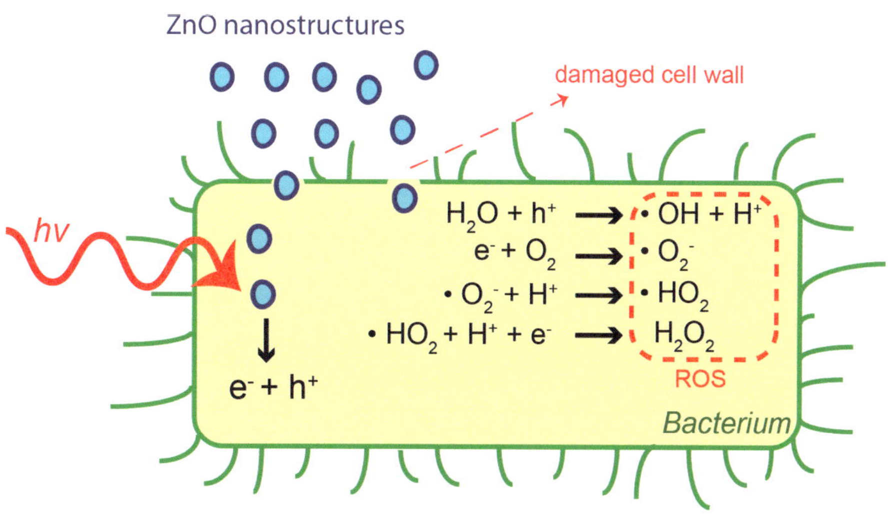

Figure 3.
ROS mechanism of ZnO nanoparticles [6].

Plant type	Zinc precursor type	Size of ZnO (nm)	Treated biomolecule	Biomedical field	Ref.
Momordica charantia	Nitrate	21	*R. microplus*, *P. humanus capitis*, *An. stephensi*, and *Cx. Quinquefasciatus*	Antiparasitic	[30]
Rosa canina	Nitrate	11	*S. typhimurium*, *S. aureus*, *L. monocytogenes*, and *E. coli*	Antibacterial	[31]
Ulva fasciata	Chloride	16	*B. cereus*, *S. aureus*, *S. thermophilus*, *E. coli*, and *P. aeruginosa*	Antibacterial	[32]
Eclipta alba	Acetate	6	*E. coli*	Antibacterial	[33]
Terminalia arjuna	Acetate	21	*E. coli* and *S. aureus*	Antibacterial	[34]
Stevia	Acetate	10–90	*E. coli* and *S. aureus*	Antibacterial Antiparasitic	[35]
Glycosmis pentaphylla	Acetate	32–36	*B. cereus*, *S. aureus*, *S. paratyphi*, *S. dysenteriae*, *C. albicans*, and *A. niger*	Antibacterial	[36]
L. leschenaultiana	Acetate	—	*L. sericata*	Antiparasitic	[37]
Adhatoda vasica	Acetate	10–12	*S. epidermidis*, *E. coli*, *P. aeruginosa*, *T. rubrum*, and *M. audouinii*	Antimicrobial	[38]
Vitex negundo	Nitrate	38	Human Serum Albumin	Protein binding	[39]
Anchusa italica	Acetate	8–14	*B. megaterium*, *S. aureus*, *E. coli*, *S. typhimurium*	Antimicrobial	[40]
Jacaranda mimosifolia	Gluconate	2–4	*E. faecium* and *E. coli*	Antibacterial	[41]
Heritiera fomes and Sonneratia apetala	Chloride	40–50	*S. aureus*, *S. flexneri*, *V. cholera*, *S. epidermidis*, *B. subtilis*, and *E. coli*	Antibacterial Anti-inflammatory	[42]
Nyctanthes arbor-tristis	Acetate	12–32	*A. alternata*, *A. niger*, *B. cinerea*, *F. oxysporum*, and *P. expansum*	Antifungal	[43]
Nocardiopsis sp	Nitrate	500	*E. coli* and *P. mirabilis*	Antibacterial	[44]
Ceropegia candelabrum	Nitrate	12–35	*S. aureus*, *B. subtilis*, *E. coli*, and *S. typhi*	Antibacterial	[45]
Pongamia pinnata	Acetate	21	*C. maculatus*	Anti-pesticide	[46]
Capsicum annuum	Nitrate	30–40	*E. coli* and *S. aureus*	Antibacterial	[47]
Euphorbia petiolata	Nitrate	—	*E. coli*	Antibacterial	[48]
Tradescantia pallida	Acetate	25	HeLa cervical cancer cell	Anticancer	[49]
Punica granatum	Nitrate	5	*P. vulgaris*, *E. coli*, and *S. aureus*	Antibacterial	[50]
Swertia chirayita	Nitrate	5	*S. enterica*, *E. coli*, and *S. aureus*	Antibacterial	[51]

Plant type	Zinc precursor type	Size of ZnO (nm)	Treated biomolecule	Biomedical field	Ref.
Vitex negundo	Nitrate	38	*E. coli* and *S. aureus*	Antibacterial	[52]
Aegle marmelos	Nitrate	20	*P. aeruginosa*, *E. coli*, *S. aureus*, *A. niger*, and *F. solani*	Antimicrobial	[53]
Azadirachta indica	Acetate	9–25	*S. pyogenes*, *E. coli*, and *S. aureus*	Antibacterial	[54]
Vaccinium arctostaphylos	Nitrate	15	Lipid, insulin, and fasting blood sugar	Antidiabetic	[55]
Coffee powder	Acetate	25–30	Proteinase K	Enzymatic	[56]
Ziziphus nummularia	Nitrate	17	HeLa cancer cell and *Candida* spp.	Anticancer Antifungal	[57]
Chelidonium majus	Nitrate	10	*P. aeruginosa*, *E. coli*, *S. aureus*, *C. albicans*, *A. niger*, and *T. rubrum*	Antibacterial	[58]
Prunus yedoensis Matsumura	Nitrate	10–40	*B. linens* and *S. epidermidis*	Antibacterial	[59]
Sechium edule	Acetate	36	*B. subtilis* and *K. pneumoniae*	Antibacterial	[60]
Catharanthus roseus	Acetate	50–90	*S. aureus*, *S. pyogenes*, *B. cereus*, *P. aeruginosa*, *P. mirabilis*, and *E. coli*	Antibacterial	[61]
Albizia saman	Nitrate	15–80	*D. indica*	Genotoxicity	[62]
Parthenium hysterophorus	Nitrate	16–45	*B. subtilis*, *S. aureus*, *K. pneumonia*, *E. coli*, and *Enterobacteraerogens*	Antibacterial	[63]
Sarcopoterium spinosum	Acetate	26–115	*E. coli*, *P. aeruginosa*, *S. aureus*, *S. pneumoniae*, *B. subtilis*, *E. faecalis*, *C. glabrata*, and *C. albicans*	Antibacterial	[64]
Ruta graveolens	Nitrate	28	*K. aerogenes*, *P. aeruginosa*, *E. coli*, and *S. aureus*	Antibacterial	[65]
Pichia kudriavzevii	Acetate	10–61	*B. subtilis*, *S. epidermidis*, *S. aureus*, *E. coli*, and *S. marcescens*	Antibacterial	[66]
Azadirachta indica	Sulfate	84	*Xanthomonas oryzae pv. oryzae*	Antibacterial	[67]
Coptidis Rhizoma	Nitrate	3–25	*B. megaterium*, *B. cereus*, *E. coli*, and *B. pumilus*	Antibacterial	[68]
Polygala tenuifolia	Nitrate	33–73	RAW 264.7 cells	Antibacterial	[69]
Sargassum muticum	Acetate	10–15	WEHI-3 cell	Anticancer	[70]
Scadoxus multiflorus	Acetate	31	*Aedes aegypti*, *A. niger*, and *A. flavus*	Antifungal	[71]
Eclipta prostrata	Nitrate	29	Hep-G2 cell	Anticancer	[72]
Azadirachta indica	Nitrate	30–60	*E. coli* and *S. aureus*	Antibacterial	[73]

Plant type	Zinc precursor type	Size of ZnO (nm)	Treated biomolecule	Biomedical field	Ref.
Calotropis gigantea	Acetate	25	DNA	Anticancer	[74]
Andrographis paniculata	Nitrate	57	α-Amylase and BSA	Antioxidant	[75]
Tamarindus indica, *Moringa oleifera*	Nitrate	27–54	α-Amylase and α-glucosidase	Antidiabetic Antioxidant	[76]
Bauhinia tomentosa	Sulfate	22–94	*B. subtilis*, *S. aureus*, *E. coli*, and *P. aeruginosa*	Antibacterial	[77]
Egg albumen	Acetate	16	*C. albicans*	Anticandidal	[78]
Moringa oleifera	Acetate	40–45	*A. saloni*, *S. rolfii*, *S. aureus*, and *E. coli*	Antibacterial	[79]
Cassia fistula	Nitrate	5–15	*K. aerogenes*, *E. coli*, *P. desmolyticum*, and *S. aureus*	Antibacterial	[80]

Table 1.
Biomedical activities of plant extract-based synthesized ZnO nanoparticles.

Sathishkumar et al. synthesized ZnO nanoflakes using *Couroupita guianensis* Aubl leaf extract and demonstrated the bactericidal activity against various types of bacteria. They reported that the main constituent of the extract, phenol, reduced the zinc acetate precursor into ZnO nanostructures. In history, Aloe (*Aloe barbadensis* Miller) extract has been used in therapeutic applications due to its antifungal, antidiabetic, anticancer, and antibacterial properties. Ali et al. also synthesized small-sized ZnO nanoparticles using *Aloe vera* extract and showed their antibacterial activity and cell damage of *Escherichia coli* and MRSA cells before and after ZnO nanoparticle treatment [7]. On the other hand, Gunalan et al. synthesized ZnO nanoparticles with *Aloe vera* extract and compared their results with the chemically synthesized ZnO nanoparticles. The results proved the enhanced antibacterial activity against various pathogens, and variation in the particle size is responsible for the significant bactericidal activity [8]. The effect of *Aloe vera* extract on the synthesis of ZnO nanoparticles and their antibacterial activity was investigated. Moreover, the antioxidant activity of the particles was evaluated by using five different free radical scavenging assays and the anticancer activity was tested against three cancerous cell lines [9].

In a very recent work, Zare et al. presented the effect of *Cuminum cyminum* leaf extract on the synthesis of ZnO nanoparticles by using zinc nitrate precursors. The resulting nanoparticle diameter is around 7 nm, and the nanoparticles show high sensitivity to Gram-negative bacteria [10]. The nanoparticle formation of zinc nitrate precursors has been investigated by using several types of plant extracts such *as Limonia acidissima* [11, 12], *Cochlospermum religiosum* [13], *Tabernaemontana divaricata* [14], *Conyza Canadensis*, *Citrus maxima* [15], *Aristolochia indica* [16], *Echinacea* [17], *Mentha* [18, 19], *Salvadora oleoides* [20], *Boswellia ovalifoliolata* [21], and *Costus pictus* [22]. The synthesized ZnO nanostructures have shown an enhanced antibacterial effect for a broad spectrum of bacterial cultures (see **Table 1**). Zinc acetate precursor is another choice for the plant-based ZnO nanoparticle synthesis, and Santoshkumar et al. synthesized ZnO nanoparticles using *Passiflora caerulea* extract against urinary tract infection pathogens. The ZnO nanostructures have a particle size of around 37 nm and show good zone of inhibition to various pathogens [23]. *Hibiscus sabdariffa* [24] and *Acalypha indica* leaves [25] were used in ZnO nanoparticle synthesis, and the nanoparticles showed an enhanced antibacterial activity against *E. coli* and *S. aureus.*

ZnO nanoparticles are currently under investigation due to their utilization in cancer treatment and diagnostic applications [26]. Since the treatment of cancer by chemotherapy is limited because of the adverse effect of tumor drugs and drug resistance by cancer cells, natural plant-based drug researches have focused on overcoming these limitations. Vijayakumar et al. investigated the anticancer activity of ZnO nanoparticles, which were synthesized by *Laurus nobilis* extract-mediated synthesis. The nanoparticles showed anti-lung cancer activity against human A549 lung cancer cells [27]. Toxicity is an important parameter for the in-vivo and in-vitro activity of nanoparticles because some nanoparticles can generate free radicals even under dark conditions. *Anacardium occidentale* extract was used in the synthesis of ZnO nanoparticles, and the resulting nanostructures exhibited concentration-dependent cytotoxicity against pancreatic cancer cells [28]. Yuvakkumar et al. utilized the Rambutan peels (*Nephelium lappaceum*) for the synthesis of ZnO nanoparticles and explored the effect of these particles on HepG2 liver cancer cells [29].

On the other hand, zinc acts as an actuator for several enzymes, and blood sugar regulation is significantly affected in the presence of zinc element. Thus, the enzymatic and anti-diabetic activity of ZnO nanoparticles must be mentioned in their biomedical applications. Bayrami et al. synthesized ZnO nanoparticles by using *Vaccinium arctostaphylos* extract via a microwave-assisted method. The biosynthesized ZnO nanoparticles showed an enhanced efficiency for the treatment of diabetic problems and reduced the fasting blood glucose level effectively [55]. Rehana et al. demonstrated the antidiabetic activity of ZnO nanoparticles by using several types of plant extracts. The results showed that *Tamarindus indica* extract-based ZnO nanoparticles exhibited enhanced activity for α-amylase and α-glucosidase due to the presence of amino acids in the plant extract [76]. As an environmentally benign material, coffee powder extract was utilized in the biosynthesis of ZnO nanoparticles. Koupaei et al. studied the reduced effect of ZnO nanoparticles on proteinase K activity [56].

2.3 Photocatalytic application

Photocatalytic degradation of organic pollutants is a promising approach for the removal of dyes in wastewaters. ZnO nanoparticles have been involved in photocatalytic applications due to their optical and electronic properties (**Table 2**). When the ZnO nanoparticles are irradiated with UV light, valence band electrons are excited to the conduction band, which leaves holes behind. Then the generated holes create hydroxyl radicals by oxidizing H_2O and OH^- and the excited electrons are captured by oxygen in the air. The resulting anionic radicals are highly reactive and degrade the organic dyes in to carbon dioxide and water (**Figure 4**).

Nava et al. addressed the effect of different amounts of *Camellia sinensis* extract on the synthesis of ZnO nanoparticles. The synthesized nanoparticles were studied in photocatalytic degradation of methylene blue (MB) dye where the nanoparticles presented MB degradation over 84% in 120 min [81]. In another study, *Parkia roxburghii* extracts have been used for the synthesis of ZnO nanoparticles, and they were found to be efficient in degradation with nearly 98% efficiency in 8 min for both MB and Rhodamine B (RhB) dyes [82]. The degradation of Congo Red dye has

Plant type	Zinc precursor type	Size of ZnO (nm)	Treated analyte	Efficiency (%) – time (min)	Ref.
Lycopersicon esculentum	Nitrate	9–20	MB	97 – 180	[86]
Monsonia burkeana	Chloride	20	MB	48 – 45	[87]
Ulva lactuca	Acetate	10–50	MB	90 – 120	[88]
Conyza canadensis	Nitrate	—	MO MB	94 – 45 85 – 20	[89]
Allium sativum	Nitrate	14–70	MB	100 – 180	[90]
Garcinia mangostana	Nitrate	21	MB	99 – 180	[91]
Plectranthus amboinicus	Nitrate	50–180	MR	92 – 180	[92]
Calotropis procera	Nitrate	15–25	MO	81 – 100	[93]
Citrus paradisi	Sulfate	12–72	MB	56 – 360	[94]
Lantana camara	Acetate	340–520	MB RhB	92 – 25 75 – 40	[95]
Chlamydomonas reinhardtii	Acetate	13	MO	92 – 120	[96]
Lycopersicon esculentum	Nitrate	7–20	MB	97 – 150	[97]
Corymbia citriodora	Nitrate	64	MB	83 – 90	[98]
Catharanthus roseus	Acetate	38	PR	100 – 480	[99]

Table 2.
Photocatalytic activities of plant-extract based synthesized ZnO nanoparticles (MB = methylene blue, MO = methyl orange, MR = methyl red, RhB = Rhodamine B, and PR = phenol red).

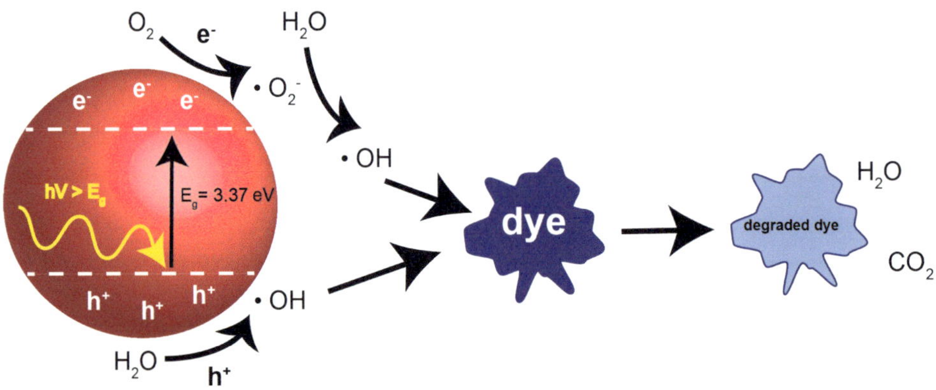

Figure 4.
Schematic diagram of dye degradation by ZnO nanostructures.

also been studied for the ZnO photocatalysis applications. Prasad et al. and Vidya et al. studied the degradation of Congo Red in aqueous solutions by ZnO nanoparticles. The dye was degraded with 90% efficiency in 35 and 60 min by using lemon juice [83] and *Artocarpus heterophyllus* [84] extracts in the nanoparticle synthesis, respectively. The aqueous leaf extract of *Coriandrum sativum* was used for the nanoparticle synthesis and the resulting materials have been used for the photocatalytic degradation of anthracene with 96% efficiency in 240 min [85].

2.4 Adsorption/sensing application

Heavy metal ion pollutants in wastewaters create a problem worldwide because of their serious effects on both human health and environment. ZnO nanostructures have also been used as an adsorbent material due to their low toxicity and became more effective than the other adsorbent materials [100]. The plant-based synthesis of ZnO nanostructures enhances metal ion adsorption capacity due to the chemical interactions between ions and functional groups of plant extracts. Fazlzadeh et al. synthesized ZnO nanoparticle-loaded activated carbon by using *Peganum harmala* for the removal of chromium from aqueous systems. *Peganum harmala* acted as a stabilizing agent and enhanced the chromium uptake up to 68.48 mg g^{-1} [101]. The lead ion removal was studied by Azizi et al. using *Zerumbone* extract-mediated ZnO nanoparticles. They reported that the lead ion adsorption capacity reached up to 19.65 mg g^{-1} [100]. Sensing application of ZnO nanostructures is another field where the glucose sensing mechanism of ZnO nanostructures was studied by Muthuchamy et al. They fabricated a glucose sensor using ZnO nanoparticles and peach juice as a carbon source. The fabricated ZnO sensor showed high sensitivity (231.7 μA mM^{-1} cm^{-2}) and low detection limit (6.3 μM) [102]. Sharma et al. also studied silymarin detection capability of ZnO nanostructures by using *Carica papaya* extract. The results showed that ZnO-modified sensors have 2-fold greater electrochemical signals than the neat ones [103].

2.5 Microorganism-based ZnO nanostructures

Synthesis of ZnO nanostructures by using microorganisms has gained considerable interest, and numerous microorganisms can be utilized for their synthesis. Bacteria, fungus, and algae are the possible microorganisms in a green synthesis of ZnO nanostructures. Because of their easy genetic manipulation and easy handling, bacteria are preferred microorganisms [104]. Jayaseelan et al. used *Aeromonas hydrophila* bacteria as green capping agent, and the synthesized ZnO nanoparticles showed antibacterial activity against *Pseudomonas aeruginosa* and *Aspergillus flavus* [104]. In another study, *Pseudomonas aeruginosa* was used as a capping agent in ZnO nanoparticle synthesis, and the resulting particles demonstrated antioxidant activity [105]. Kundu et al. used *Rhodococcus pyridinivorans* as metabolically versatile Actinobacteria in the fabrication of self-cleaning, UV-blocking, and antibacterial textile fabrics with ZnO nanoparticles. Besides, the ZnO nanoparticles showed photocatalytic activity against malachite green and anticancer activity against HT-29 cancerous cells [106]. Tripathi et al. synthesized ZnO nanoflowers by using *Bacillus licheniformis* and assessed their photocatalytic activity against methylene blue [107]. However, bacterial utilization in ZnO green synthesis could be somewhat problematic because of the uncontrolled growth of bacteria and unavoidable contaminations [108].

Fungus-based green synthesis of ZnO nanoparticles is generally preferred over the bacteria based synthesis because of their large-scale production and better tolerance property [109]. Raliya et al. synthesized ZnO nanoparticles via an environmental method by using *Aspergillus fumigatus* as a stabilizing agent and investigated the effect of ZnO nanoparticles on phosphorus mobilizing enzymes in rhizosphere and gum contents in clusterbean grains [110]. In another study, ureolytic bacterium (*Serratia ureilytica*)-mediated green synthesis of ZnO nanoparticles was reported. The cotton fabrics were coated with the synthesized nanoparticles and their killing efficiency against *S. aureus* and *E. coli* bacteria was revealed [111].

Algae are the members of a diverse group of aquatic photosynthetic organisms and they have been utilized sometimes in the synthesis of ZnO nanostructures. *Sargassum muticum* extract, which is a brown marine macroalga, was used in the biosynthesis of ZnO nanoparticles [112]. Nagajaran et al. used the seaweed extracts of green *Caulerpa peltata*, red *Hypnea Valencia* and *brown Sargassum myriocystum* in the synthesis of ZnO nanoparticles. The results revealed that among three seaweeds, only *S. myriocystum* could stabilize and reduce ZnO nanoparticles of size 36 nm. Also, the nanoparticles showed antimicrobial activity against a wide spectrum of bacterial cultures [109].

2.6 Biomolecule-based ZnO nanostructures

Synthesis of ZnO nanoparticles with controlled morphologies and using environmentally friendly chemicals could be possible in biomolecule-based synthesis routes, and utilization of amino acids, polysaccharides, gums, and enzymes is highly preferable. Gharagozlou et al. synthesized ZnO nanoparticles by using alanine amino acid, and a Schiff base complex was obtained at the end of the study [113]. Bovine skin gelatin has also been used in the synthesis of ZnO nanostructures, and Alnarabiji et al. demonstrated the environmentally friendly synthesis route for ZnO nanoparticles [114]. Arabic gum and *gum tragacanth*-based green synthesis of ZnO nanoparticles have been demonstrated by Fardood et al. [115] and Daraoudi et al. [116], respectively. Thus, they demonstrated an alternative method for the synthesis of ZnO nanoparticles instead of conventional ZnO reduction methods by using hazardous polymers or surfactants [115, 116]. Casein is another biomolecule that can be used as a capping and reducing agent in the ZnO nanoparticle synthesis. Somu et al. synthesized ZnO nanoparticles, which show heavy metal ion adsorption, dye adsorption, and antibacterial activity in wastewater treatment at the same time [117]. The resulting nanoparticles effectively remove Cd(III), Pd(II), and Co(II) ions, methylene blue, and Congo red dye from the wastewaters. Also, they demonstrated high antibacterial activity against *E. coli* cultures. Subramanian et al. synthesized ZnO nanoflowers, which comprise nanorods and ellipsoids as subunits, by using L-lysine as a capping and precipitating agent [118].

3. Synthesis of ZnO nanostructures using nontoxic chemicals

Combating the major drawbacks of common ZnO nanostructure synthesis methods, mainly identified as the generation of pollutants, toxic materials, and side products during reactions, green chemical techniques using only nontoxic and biologically compatible materials were developed. Gharagozlou et al. [113] reported a novel method to synthesize ZnO nanoparticles without any pollutant or combustible side product in the process. Water was used as a solvent with a biologically compatible nitrogen source, amino acid instead of toxic amines, alanine and sodium salicylaldehyde-5-sulfonate, and zinc acetate to prepare the zinc Schiff-base complex and then subsequently heated to obtain ZnO nanoparticles. This work showed that the solid-state decomposition process applied at moderate temperatures has yielded nanoparticles ranging from 5 to 110 nm with fewer defects yet interestingly high crystallinity.

Biomimetic and bioinspired synthesis has also been regarded among the most attractive strategies in fabricating novel functional materials, and biological materials like eggshell membrane, oyster shells, nacre, diatoms, cuttlefish bone, DNA chains, and sea urchin spines have been actually employed as templates or bioreactive substrates. Silk fibroin fibers (SFFs) extracted from silkworm *Bombyx mori* cocoons were used for their capping and directing functions to control the morphology of ZnO crystals. Acting at the same time, zinc ions are anchored on the SFF and in-situ react with OH^- generating ZnO nanoparticles. It was observed that several petals composed of ZnO nanoparticles form ZnO flowers and bestrewed the SFF substrates due to the electron-donating groups (amino and carboxyl groups) contained in SFF [119]. Another ZnO nanoparticle synthesis method through biological roots such as the use of natural biopolymers was to be presented as more cost-effective compared to both physical and chemical methods available [120]. A recent work has focused on integrating ZnO nanoparticles with biopolymers that are excellent vehicles for cross-linking molecules. Thereby, collagen was used for zinc acetate in basic solution, to give a precipitate that was processed and thermally heated at 350°C giving birth to ZnO nanoparticles, yielding an interesting antibiofilm, anti-cancer, and ecotoxicity material [121].

Extrusion dripping is another novel technique using environmentally friendly, cost-effective, degradable, and renewable biomass materials. Generally, the process yields monodispersed spherical particles by controlled dripping of working solution into a biopolymer solution after extruding it through a narrow tube, thanks to the effect of viscous-surface tension forces and impact-drag forces that help to preserve the spherical shape of the drop [122]. Goes et al. have reported that spherical uniform sized ZnO nanoparticles were obtained by dropwise addition of alginate solution to zinc nitrate solution under a long slow magnetic stirring to ensure the ion-exchange to happen and stabilize ZnO nanoparticles; the heated ZnO outcome was used to fabricate a polymeric nanocrystalline microfilm, exhibiting interesting photodegradation results, as ZnO on the surface is likely to accept photons and generate holes promoting the oxidative decomposition of the dye [123].

Chemical bath deposition and soft-template sol-gel methods are two wet organic solvent-free routes studied by a group led by Leone et al [124]. to obtain a nanostructured ZnO employed as a reservoir of clotrimazole for pharmaceutical purposes. Identifying the synthesis of carriers and active pharmaceutical ingredient loading as the main steps in which the waste of organic solvents occur, ZnO nanostructures have been introduced as green alternative carriers for their intrinsic biological properties, low toxicity, and high biocompatibility [125]. For the chemical bath deposition approach, a nanosheet-like zinc carbonate hydroxide hydrate was transformed into ZnO using solutions containing urea and different zinc salts [126]. As for the sol-gel method, pluronic F127 was used as a soft template forming an opalescent solution with zinc acetate in water, and then dried and calcined at 500°C to sacrifice the template and obtain the ZnO nanostructure [124].

In our previous work, we successfully enhanced the antibacterial activity of ZnO nanowires by modifying the cooling route. Zinc acetate was calcined in a muffle oven, followed by a rapid cooling; the three resulting samples were compared to a free cooled batch synthesized under the same conditions, revealing noticeable effects on ZnO nanowire morphology in addition to the improvement of surface area due to the limiting time for crystallite growth [127].

3.1 Ionic liquid as solvents for ZnO nanostructures

Ionic liquids (ILs) are an area of chemistry, which has received important attention in both academia and industry, because of the cost effectiveness coupled with being environmentally friendly [128]. ILs usually act as solvents and reactants as well as templates for inorganic nanomaterial synthesis and scavenging agents [129]. As a subdivision, room temperature ionic liquids (RTILs) are particularly doted of special considerations as nontoxic solvents with a wide liquid temperature range, remarkable chemical stability, negligible vapor pressures, and high fire resistance [130]. ILs have been great templates for the synthesis of nanomaterials as it was shown that only by modifying the structure of their cations or anions, it is possible to alter their properties in order to control the size, morphology, and thus the properties of nanomaterials [131]. Sabbaghan et al. have synthesized different morphologies of ZnO nanostructures using zinc acetate as the metal source in a basic media to react with different symmetrical imidazolium-based ILs, yielding nanoparticles, nanoparticle-like, spherical-like, nanosheet in different sizes ranging from 16 to 30 nm and different band gaps between 2.98 and 3.17 eV, demonstrating through this work the relation morphology-IL [128].

ZnO nanostructures with nanosheet morphology have been successfully fabricated in another work by refluxing the mixture of zinc acetate and the ionic liquid in water according to Menshutkin reaction [132]. A comparative study has shown that when zinc is used with ILs as a template, ZnO nanoparticles with smaller crystallite size were formed compared to the yield without ILs. This work revealed as well that the template and the pH control the direction of growth of ZnO crystals and the shape of nanomaterials obtained. The final band gap values of ZnO with different morphology ranged around 2.88–3.16 eV [133]. Alammar et al. have studied the effect of five different ILs on ZnO morphologies, and claimed that the habitus and morphology come as the system naturally tends to reduce the total surface energy during formation; it is to note that the anion of the IL is proposed to be interacting with the ZnO surface during the growth [134]. Yet, the best performance was for ZnO nanoparticles that are obtained by use of IL with a long alkyl chain, reaching 95% in 9 h for methyl orange decomposition, proposing that along with high surface area, oxygen vacancies and polar plans that act like electron traps are the main factors for such interesting photocatalytic activity [135, 136]. It was also reported by Amde et al. [137] that common techniques for the determination of fungicide concentration in water are usually non-environmentally friendly, organic solvent, and time consuming; the group has prepared ZnO nanofluids by a green two-step method, dispersing the as-synthesized sol-gel ZnO nanoparticles in 1-hexyl-3-methylimidazolium hexafluorophosphate, hand shaking it to attain a homogeneous distribution, then sonicating it to break NPs clusters. The as-prepared ZnO nanofluids are applied in a modified, simple, versatile, and inexpensive liquid-liquid microextraction technique; this technique, called single drop microextraction not only reduces the amount of extraction solvent radically but also offers other functionalities such as high enrichment factor, different extraction modes, and full automation of the process [138]. It is proclaimed that the preferences of ZnO-based nanofluids for the investigation were driven from the fact that ZnO dotes on surface charges that enable to form stable suspensions, unlike many metallic nanofluids, and without the need of any additional stabilizer intervention.

4. Processes of synthesis of ZnO nanostructures

ZnO nanostructures can be synthesized by following different approaches, and each has its own distinctive advantages and downsides. In this section, we report some interesting methods and findings having the common main aim to avoid drawbacks like the use of toxic reagents, promoter, and stabilizer organic additives, lowering the reaction time, as well as high temperature and pressure. These methods have plenty of scopes to provide both qualitative and quantitative support for nanosized ZnO synthesis along with being simple, fast, efficient, and convenient.

4.1 Hydrothermal synthesis of ZnO nanostructures

Hydrothermal method has gained particular interest as an efficient method for high quality and mass production of ZnO nanostructures [139]; it is indeed environmentally friendly as there is no need to control pH and subsequently no release of unwanted by-products. A work led by Guo et al. [140] reported a controllable hydrothermal synthesis of ZnO nanorods reacting with zinc carbonate hydroxide hydrate powder and H_2O_2 at various temperatures for different periods of time. The group claims that the formation mechanism of ZnO starts by the formation of ZnO_2 when subjecting it to hydrothermal treatment at 170°C for more time (3–6 h), and then the thermally unstable ZnO_2 would decompose into ZnO and O_2. ZnO nanorods exhibited an optical band gap of 3.3 eV [140].

It has also been reported that flower shaped ZnO nanoparticles were synthesized by hydrothermal method, where zinc nitrate and hexamethylenetetramine solutions were prepared separately in double distilled water, while NaOH solution was added dropwise to adjust the pH to 10. The obtained milky solution was refluxed at 80°C for 7 h, washed, and dried. XRD asserted the formation of hexagonal crystal structure ZnO that had a flower-shape, formed by agglomeration during the hydrothermal process. Dynamic light scattering (DLS) data have affirmed the average diameter of ZnO between 600 and 800 nm. The effect of different pH values from 2 to 10 on the removal performance of ZnO has been studied and the results showed a maximum increase of 80% removal efficiency when pH reached 6; increasing the temperature of the process also improved the removal efficiency from 68 to 97%. The contact time had a sharp rise at the value of 15 min of initiation of the experiment, and higher stirring had an enhancing effect as well. Jamal Al-Sabahi and his group have treated for the first time the degradation of HPAM polymer in oil produced water with supported ZnO nanorods synthesized via a microwave-assisted hydrothermal method in an aqueous solution [141–143]. Placing a prepared microscope glass substrate (25 mm × 75 mm) on a hotplate (350°C), 10 mM zinc acetate dihydrate solution is sprayed, and then the plate is immersed in an equimolar solution of zinc nitrate hexahydrate and hexamethylenetetramine, then heated in a domestic microwave oven for 45 min, and then cooled down for 15 min; the produced ZnO nanorod-covered substrate was afterward annealed in air at 350°C for 1 h. The morphology was of a typical ZnO array and the average length was about 4 μm while the average diameter reached around 95 nm [144].

Singh et al. have adopted a cost-effective and environmentally friendly method to fabricate a 3D self-assembled wool ball like spherical ZnO with high porosity by combining a urea-glycerol assisted hydrothermal approach with calcination under air atmosphere [145], where hydrated zinc carbonate was synthesized hydrothermally in a Teflon-lined stainless steel autoclave reacting zinc nitrate with urea in a triplet (glycerol, ethyl alcohol, and water (7:7:10)) solvent system. After drying the outcome, the white powder intermediate product was calcined at 450°C for 3 h to get hierarchical 3D porous ZnO. The group carried out a series of experiments to investigate the effect of synthesis parameters over the morphology, while 7 h and 140°C were about the optimum duration of synthesis and temperature to get the best ZnO with W-ball like spherical morphology. This ZnO photocatalyst has shown 98% of highly toxic Rhodamine B degradation in 60 min of UV photolysis catalysts at a pH of 4.

A green hydrothermal method was used recently to fabricate ZnO nanorods without any organic solvent or surfactant, by starting with ZnO powder and H_2O_2 aqueous solution in the sealed autoclave. Then the precipitates were washed and dried. In this work, Lam et al. reported as well that the hydrothermal treatment of ZnO at 100°C promotes the slow conversion of ZnO_2 to ZnO and O_2 without using any toxic reactant, nor releasing any pollutant by-product. XRD indicated the high purity of the ZnO wurtzite phase [146]. This environmentally friendly method has generated highly performing ZnO nanorods that completely degrade the resorcinol in aqueous solution after 120 min. The same group has used a similar hydrothermal method to fabricate ZnO nanotubes (NTs) with a diameter of around 10 nm, a wall thickness of 3.5 nm, and average lengths of up to 200 nm by scrolling of the ZnO_2 layer nanosheet, which transforms to ZnO NTs with a symmetrical layer in both shell-tube structures. XRD asserted the hexagonal phase and then confirmed the purity of ZnO NTs that showed a ferromagnetic behavior because of the grain boundaries and developed free surfaces [147]; the band gap energy was measured to be around 3.21 eV, and the degradation of methylparaben over the surface of ZnO NTs had an efficiency of 87.6% in 105 min [148].

4.2 Microwave-assisted synthesis of ZnO nanostructures

Microwave-assisted sol-gel synthesis is based on subjecting samples to frequencies ranging from 300 MHz to 300 GHz [149]; besides its high selectivity of specific morphologies, dramatic reduction of reaction time (minutes), and remarkable increase of product yield, it generates localized superheating at the reaction sites promoting metal ion reduction in the solution [150]. It presents a promising green method for metal oxide nanostructures production. Azizi et al. [151] have worked on a green microwave-assisted combustion approach to synthesize ZnO-nanoparticles, presenting combustion as a fast, low cost, homogenous and highly pure outcome, as well a distinguished surface area at low temperature. Using fruit, seed, and pulp extracts of *Citrullus colocynthis* (L.) as biofuels with zinc nitrate as the zinc source, an in vitro cytotoxicity study has been made showing that smaller nanoparticles were more efficient penetrating in the cells membranes, while the optical band gap increased with the rear of the particle size from 3.25 to 3.40 eV. An interesting nontoxic and eco-friendly single-step, and green synthesis method of ZnO nanoparticles with excellent reproducibility was reported using coffee powder extract as a reducing material under microwave heating at 540 W for 5 min and the precipitate was dried in a hot air oven. SEM asserted a size of 80–120 nm for the nanoparticles. Afterward, a thin nanocomposite film was prepared using the as-prepared ZnO nanoparticles with natural graphite powder. The nanocomposite film showed a remarkable photovoltaic efficiency of 3.12% [152]. ZnO sub-micrometer particles and nanowires were synthesized by microwave assisted sol-gel reaction. Zinc acetate and *N*,*N*-dimethylacetamide were stirred into a beaker, and then, the solution was cooled rapidly to 15°C. After only a couple of minutes of the microwave, a white suspension was obtained. The average diameter of the particles prepared given by DLS analyses was around 275–352 nm [153]. Salari et al. reported a microwave-assisted synthesis of biogenic nanoparticles using *Lavandula vera* leaf extract as a reducing agent in the presence of zinc sulfate; the method led to simple and fast formation of microstructures that exhibited high antioxidant cytotoxic activity [154]. In a separate work, vertically aligned ZnO nanorods were grown at 90°C on a Si substrate by microwave synthesis and compared with nanorods made by the traditional heated waterbath method changing the pH from 10.07 to 10.9. The microwave synthesis was performed at a power of 100 W [155]. The same group observed that the increase of ammonia led in both methods to sparser and longer nanorods with the larger diameter, as well as an increase of oxygen percentages in the samples. The microwave synthesized samples exhibit a uniform distribution of nanorods as well as a better crystalline structure with fewer defects than the heated water bath-grown samples, which can be beneficial for band-edge transition optoelectronic devices.

4.3 Sonochemical synthesis of ZnO nanostructures

Microwave synthesis and ultrasound sonication have proven numerous advantages and actually become amid the most frequently sought nanosized material synthesis methods [156, 157]. Zak et al. have used ultrasonication to synthesize ZnO nanostructures at room temperature without any specific conditions or organic solvents, starting from an as-prepared zinc solution where zinc acetate was dissolved in ammonia solution, and sodium hydroxide was dropped in the solution. Deionized water was wisely added till attaining a concentration of 1 M zinc. The ultrasonication performed at different durations was sufficient to stimulate the formation of nanostructures. XRD asserted the presence of hexagonal pure ZnO and the band gap energies estimated by UV-Vis spectra are 3.3, 3.22, and 3.2 eV for ZnO seeds, nanorods, and nanoflowers, respectively [158]. Ultrasound was conducted as well for the synthesis of different ZnO nanostructures without any organic solvents, surfactants, or templating agents; zinc acetate and sodium hydroxide in ionic liquids (ILs) are reported to be a green, fast, and effective, yet highly selective route to 0D, 1D, and 2D nanostructures of ZnO [134]. A facile calcination-free ultrasound assisted approach has been reported by Bhatte et al. involving zinc acetate as a metal source and 1, 3-propane diol as a solvent, base, stabilizer, and template for the growth of nanocrystalline ZnO [159]. The mixture of both materials has been sonicated under 22 kHz frequency for 2 h with a 5 s interval on-off pulse. After sonication, the formed ZnO was collected, washed and dried. XRD confirmed the successful formation of ZnO without any impurities [159].

4.4 Low temperature synthesis of ZnO nanostructures

For the large-scale production of pure ZnO nanocrystallites, low thermal processes are one of the most efficient methods minimizing the generated waste yet implementing sustainable processes: simple, cheap, and nonpolluting, addressing the key issues that draw much consideration in a green solid-state synthetic method, by eliminating the use of nontoxic materials and reducing energy consumption. ZnO_2 nanocrystallites were employed as the precursor for ZnO production, because of their facile preparation, the absence of unwanted by-products, and low-temperature decomposition reaction [160]. Zinc acetate and hydrogen peroxide were used first to synthesize ZnO_2 nanocrystallites hydrothermally at 100°C for 12 h in an alkaline aqueous solution; the product was subjected to 180°C in air for 12 h yielding pure ZnO phase shaped nanocrystallites of 8–10 nm and blue shift at around 350 nm in the UV-Vis spectra [161].

ZnO nanosized structures were also synthesized starting from zinc acetate and urea in a 1:1 stoichiometry, where the decomposition of urea helped the formation of ZnO. Then two different heating methods were applied: microwave hydrothermal (MH) method and waterbath heating. XRD proved the purity of the wurtzite phase for both methods, while FE-SEM showed a difference in shape regularity in favor of the MH process; thus, the MH method contributes to the production of spherical and uniform particles after a short processing time by enhancing the interface mobility and the diffusivity in the medium [162].

Raja et al. have reported a laboratory procedure based on sol-gel for the preparation of nano-ZnO particles. Zinc acetate solution was stirred at room temperature while adding sodium hydroxide until reaching a pH of 14 and the solution went through a microemulsion. The suspension obtained was transferred for thermal treatment at 180°C for 3 h, and then the white precipitate was collected, washed, centrifuged, and dried under vacuum to reveal well-shaped uniform ZnO nanoparticles of 35 nm on average [163].

Another group has proposed a simple, eco-friendly approach to synthesize ZnO nanoparticles by using carboxylic curdlan (cc) as a reducing and stabilizing agent. A solution of zinc acetate was blended with cc aqueous solution and stirred at 70°C for 6 h, and then the outcome was freeze-dried [164]. The carboxyl group is in charge of chelating and reducing zinc ions for the sake of the formation of ZnO nanoparticles, thanks to the numerous negatively charged carboxyl groups it contains. The average diameter of cc-ZnO nanoparticles is around 58 nm exhibiting a band gap energy of 3.3 eV. Furthermore, the interaction between the as-prepared nanoparticles and bovine serum albumin (BSA) at room temperature was investigated, which suggested the formation of a certain complex revealed by a blue shift of the fluorescence peak by about 8 nm with increasing nanoparticle concentration, due to the binding of cc-ZnO nanoparticles and BSA [164].

Conflict of interest

The authors declare no conflict of interest.

Author details

Tuğba Isık[1], Mohamed Elhousseini Hilal[2] and Nesrin Horzum[3]*

1 Materials Science and Engineering Department, İzmir Institute of Technology, Turkey

2 State Key Laboratory of Advanced Technology for Materials Synthesis and Processing, Wuhan University of Technology, China

3 Department of Engineering Sciences, Izmir Katip Celebi University, Izmir, Turkey

*Address all correspondence to: nesrin.horzum.polat@ikc.edu.tr

References

[1] Agarwal H, Menon S, Kumar SV, Rajeshkumar S. Mechanistic study on antibacterial action of zinc oxide nanoparticles synthesized using green route. Chemico-Biological Interactions. 2018;**286**:60-70

[2] Vishnukumar P, Vivekanandhan S, Misra M, Mohanty AK. Recent advances and emerging opportunities in phytochemical synthesis of ZnO nanostructures. Materials Science in Semiconductor Processing. 2018;**80**:143-161

[3] Thema FT, Manikandan E, Dhlamini MS, Maaza M. Green synthesis of ZnO nanoparticles via Agathosma betulina natural extract. Materials Letters. 2015;**161**:124-127

[4] Ahmed S, Annu, Chaudhry SA, Ikram S. A review on biogenic synthesis of ZnO nanoparticles using plant extracts and microbes: A prospect towards green chemistry. Journal of Photochemistry and Photobiology B-Biology. 2017;**166**:272-284

[5] Cowan MM. Plant products as antimicrobial agents. Clinical Microbiology Reviews. 1999;**12**:564

[6] Taran M, Rad M, Alavi M. Biosynthesis of TiO_2 and ZnO nanoparticles by Halomonas elongata IBRC-M 10214 in different conditions of medium. BioImpacts: BI. 2018;**8**:81-89

[7] Ali K, Dwivedi S, Azam A, Saquib Q Al-Said MS, Alkhedhairy AA, et al. Aloe vera extract functionalized zinc oxide nanoparticles as nanoantibiotics against multi-drug resistant clinical bacterial isolates. Journal of Colloid and Interface Science. 2016;**472**:145-156

[8] Gunalan S, Sivaraj R, Rajendran V. Green synthesized ZnO nanoparticles against bacterial and fungal pathogens. Progress in Natural Science-Materials International. 2012;**22**:695-702

[9] Mahendiran D, Subash G, Selvan DA, Rehana D, Kumar RS, Rahiman AK. Biosynthesis of zinc oxide nanoparticles using plant extracts of *Aloe vera* and *Hibiscus sabdariffa*: Phytochemical, antibacterial, antioxidant and anti-proliferative studies. Bionanoscience. 2017;**7**:530-545

[10] Zare E, Pourseyedi S, Khatami M, Darezereshki E. Simple biosynthesis of zinc oxide nanoparticles using nature's source, and it's in vitro bio-activity. Journal of Molecular Structure. 2017;**1146**:96-103

[11] Patil BN, Taranath TC. *Limonia acidissima* L. leaf mediated synthesis of silver and zinc oxide nanoparticles and their antibacterial activities. Microbial Pathogenesis. 2018;**115**:227-232

[12] Taranath TC, Patil BN. *Limonia acidissima* L. leaf mediated synthesis of zinc oxide nanoparticles: A potent tool against *Mycobacterium tuberculosis*. International Journal of Mycobacteriology. 2016;**5**:197-204

[13] Mahendra C, Murali M, Manasa G, Ponnamma P, Abhilash MR, Lakshmeesha TR, et al. Antibacterial and antimitotic potential of bio-fabricated zinc oxide nanoparticles of *Cochlospermum religiosum* (L.). Microbial Pathogenesis. 2017;**110**:620-629

[14] Raja A, Ashokkumar S, Marthandam RP, Jayachandiran J, Khatiwada CP, Kaviyarasu K, et al. Eco-friendly preparation of zinc oxide nanoparticles using Tabernaemontana divaricata and its photocatalytic and antimicrobial activity. Journal of Photochemistry and Photobiology B-Biology. 2018;**181**:53-58

[15] Pavithra NS, Lingaraju K, Raghu GK, Nagaraju G. *Citrus maxima* (Pomelo) juice mediated eco-friendly synthesis of ZnO nanoparticles: Applications to photocatalytic, electrochemical sensor and antibacterial activities. Spectrochimica Acta Part A-Molecular and Biomolecular Spectroscopy. 2017;**185**:11-19

[16] Steffy K, Shanthi G, Maroky AS, Selvakumar S. Enhanced antibacterial effects of green synthesized ZnO NPs using *Aristolochia indica* against Multi-drug resistant bacterial pathogens from Diabetic Foot Ulcer. Journal of Infection and Public Health. 2018;**11**:463-471

[17] Attar A, Yapaoz MA. Biomimetic synthesis, characterization and antibacterial efficacy of ZnO and Au nanoparticles using echinacea flower extract precursor. Materials Research Express. 2018;**5**:055403

[18] Mohammadi-Aloucheh R, Habibi-Yangjeh A, Bayrami A, Latifi-Navid S, Asadi A. Green synthesis of ZnO and ZnO/CuO nanocomposites in *Mentha longifolia* leaf extract: Characterization and their application as antibacterial agents. Journal of Materials Science: Materials in Electronics. 2018;**29**:13596-13605

[19] Karakose E, Colak H. Structural, electrical, and antimicrobial characterization of green synthesized ZnO nanorods from aqueous Mentha extract. MRS Communications. 2018;**8**:577-585

[20] Padalia H, Baluja S, Chanda S. Effect of pH on size and antibacterial activity of *Salvadora oleoides* leaf extract-mediated synthesis of zinc oxide nanoparticles. Bionanoscience. 2017;**7**:40-49

[21] Supraja N, Prasad T, Krishna TG, David E. Synthesis, characterization, and evaluation of the antimicrobial efficacy of Boswellia ovalifoliolata stem bark-extract-mediated zinc oxide nanoparticles. Applied Nanoscience. 2016;**6**:581-590

[22] Suresh D, Nethravathi PC, Udayabhanu, Rajanaika H, Nagabhushana H, Sharma SC. Green synthesis and characterization of zinc oxide nanoparticle using insulin plant (*Costus pictus* D. Don) and investigation of its antimicrobial as well as anticancer activities. Materials Sciencein SemiconductorProcessing. 2015;**31**:446-454

[23] Santhoshkumar J, Venkat Kumar S, Rajeshkumar S. Synthesis of zinc oxide nanoparticles using plant leaf extract against urinary tract infection pathogen. Resource-Efficient Technologies. 2017;**3**:459-465

[24] Bala N, Saha S, Chakraborty M, Maiti M, Das S, Basu R, et al. Green synthesis of zinc oxide nanoparticles using *Hibiscus subdariffa* leaf extract: Effect of temperature on synthesis, anti-bacterial activity and anti-diabetic activity. RSC Advances. 2015;**5**:4993-5003

[25] Karthik S, Siva P, Balu KS, Suriyaprabha R, Rajendran V, Maaza M. Acalypha indica-mediated green synthesis of ZnO nanostructures under differential thermal treatment: Effect on textile coating, hydrophobicity, UV resistance, and antibacterial activity. Advanced Powder Technology. 2017;**28**:3184-3194

[26] Saranya S, Vijayaranai K, Pavithra S, Raihana N, Kumanan K. In vitro cytotoxicity of zinc oxide, iron oxide and copper nanopowders prepared by green synthesis. Toxicology Reports. 2017;**4**:427-430

[27] Vijayakumar S, Vaseeharan B, Malaikozhundan B, Shobiya M. Laurus nobilis leaf extract mediated green synthesis of ZnO nanoparticles: Characterization and biomedical

applications. Biomedicine & Pharmacotherapy. 2016;**84**:1213-1222

[28] Zhao CB, Zhang XX, Zheng YM. Biosynthesis of polyphenols functionalized ZnO nanoparticles: Characterization and their effect on human pancreatic cancer cell line. Journal of Photochemistry and Photobiology B-Biology. 2018;**183**:142-146

[29] Yuvakkumar R, Suresh J, Saravanakumar B, Nathanael AJ, Hong SI, Rajendran V. Rambutan peels promoted biomimetic synthesis of bioinspired zinc oxide nanochains for biomedical applications. Spectrochimica Acta Part A-Molecular and Biomolecular Spectroscopy. 2015;**137**:250-258

[30] Gandhi PR, Jayaseelan C, Mary RR, Mathivanan D, Suseem SR. Acaricidal, pediculicidal and larvicidal activity of synthesized ZnO nanoparticles using *Momordica charantia* leaf extract against blood feeding parasites. Experimental Parasitology. 2017;**181**:47-56

[31] Jafarirad S, Mehrabi M, Divband B, Kosari-Nasab M. Biofabrication of zinc oxide nanoparticles using fruit extract of *Rosa canina* and their toxic potential against bacteria: A mechanistic approach. Materials Science & Engineering C-Materials for Biological Applications. 2016;**59**:296-302

[32] Bhutiya PL, Mahajan MS, Rasheed MA, Pandey M, Hasan SZ, Misra N. Zinc oxide nanorod clusters deposited seaweed cellulose sheet for antimicrobial activity. International Journal of Biological Macromolecules. 2018;**112**:1264-1271

[33] Singh AK, Pal P, Gupta V, Yadav TP, Gupta V, Singh SP. Green synthesis, characterization and antimicrobial activity of zinc oxide quantum dots using &ITEclipta alba&IT. Materials Chemistry and Physics. 2018;**203**:40-48

[34] Saha R, Karthik S, Balu KS, Suriyaprabha R, Siva P, Rajendran V. Influence of the various synthesis methods on the ZnO nanoparticles property made using the bark extract of Terminalia arjuna. Materials Chemistry and Physics. 2018;**209**:208-216

[35] Khatami M, Alijani HQ Heli H, Sharifi I. Rectangular shaped zinc oxide nanoparticles: Green synthesis by Stevia and its biomedical efficiency. Ceramics International. 2018;**44**:15596-15602

[36] Vijayakumar S, Krishnakumar C, Arulmozhi P, Mahadevan S, Parameswari N. Biosynthesis, characterization and antimicrobial activities of zinc oxide nanoparticles from leaf extract of *Glycosmis pentaphylla* (Retz.) DC. Microbial Pathogenesis. 2018;**116**:44-48

[37] Banumathi B, Vaseeharan B, Malaikozhundan B, Ramasamy P, Govindarajan M, Alharbi NS, et al. Green larvicides against blowflies, *Lucilia sericata* (Diptera, Calliphoridae): Screening of seven plants used in Indian ethno-veterinary medicine and production of green-coated zinc oxide nanoparticles. Physiological and Molecular Plant Pathology. 2018;**101**:214-218

[38] Sonia S, Kumari HLJ, Ruckmani K, Sivakumar M. Antimicrobial and antioxidant potentials of biosynthesized colloidal zinc oxide nanoparticles for a fortified cold cream formulation: A potent nano cosmeceutical application. Materials Science & Engineering C-Materials for Biological Applications. 2017;**79**:581-589

[39] Ambika S, Sundrarajan M. Green biosynthesis of ZnO nanoparticles using *Vitex negundo* L. extract: Spectroscopic investigation of interaction between ZnO nanoparticles and human serum albumin. Journal of Photochemistry and Photobiology B-Biology. 2015;**149**:143-148

[40] Azizi S, Mohamad R, Bahadoran A, Bayat S, Rahim RA, Ariff A, et al. Effect of annealing temperature on antimicrobial and structural properties of bio-synthesized zinc oxide nanoparticles using flower extract of *Anchusa italica*. Journal of Photochemistry and Photobiology B-Biology. 2016;**161**:441-449

[41] Sharma D, Sabela MI, Kanchi S, Mdluli PS, Singh G, Stenstrom TA, et al. Biosynthesis of ZnO nanoparticles using Jacaranda mimosifolia flowers extract: Synergistic antibacterial activity and molecular simulated facet specific adsorption studies. Journal of Photochemistry and Photobiology B-Biology. 2016;**162**:199-207

[42] Thatoi P, Kerry RG, Gouda S, Das G, Pramanik K, Thatoi H, et al. Photo-mediated green synthesis of silver and zinc oxide nanoparticles using aqueous extracts of two mangrove plant species, Heritiera fomes and Sonneratia apetala and investigation of their biomedical applications. Journal of Photochemistry and Photobiology B-Biology. 2016;**163**:311-318

[43] Jamdagni P, Khatri P, Rana JS. Green synthesis of zinc oxide nanoparticles using flower extract of *Nyctanthes arbor-*tristis and their antifungal activity. Journal of King Saud University Science. 2018;**30**:168-175

[44] Rajivgandhi G, Maruthupandy M, Muneeswaran T, Anand M, Manoharan N. Antibiofilm activity of zinc oxide nanosheets (ZnO NSs) using Nocardiopsis sp GRG1 (KT235640) against MDR strains of gram negative *Proteus mirabilis* and *Escherichia coli*. Process Biochemistry. 2018;**67**:8-18

[45] Murali M, Mahendra C, Nagabhushan, Rajashekar N, Sudarshana MS, Raveesha KA, et al. Antibacterial and antioxidant properties of biosynthesized zinc oxide nanoparticles from *Ceropegia candelabrum* L.—An endemic species. Spectrochimica Acta Part A-Molecular and Biomolecular Spectroscopy. 2017;**179**:104-109

[46] Malaikozhundan B, Vinodhini J. Nanopesticidal effects of *Pongamia pinnata* leaf extract coated zinc oxide nanoparticle against the *Pulse beetle*, *Callosobruchus maculatus*. Materials Today Communications. 2018;**14**:106-115

[47] Lalithamba HS, Raghavendra M, Uma K, Yatish KV, Mousumi D, Nagendra G. *Capsicum annuum* fruit extract: A novel reducing agent for the green synthesis of ZnO nanoparticles and their multifunctional applications. Acta Chimica Slovenica. 2018;**65**:354-364

[48] Mohammadi C, Mahmud SA, Abdulla SM, Mirzaei Y. Green synthesis of ZnO nanoparticles using the aqueous extract of *Euphorbia petiolata* and study of its stability and antibacterial properties. Moroccan Journal of Chemistry. 2017;**5**:476-484

[49] Li CY, Zhang ZC, Mao JY, Shi LF, Zheng Y, Quan JL. Preparation of *Tradescantia pallida*-mediated zinc oxide nanoparticles and their activity against cervical cancer cell lines. Tropical Journal of Pharmaceutical Research. 2017;**16**:494-500

[50] Fuku X, Diallo A, Maaza M. Nanoscaled electrocatalytic optically modulated ZnO nanoparticles through green process of *Punica granatum* L. and their antibacterial activities. International Journal of Electrochemistry. 2016;**2016**:4682967

[51] Akhter SMH, Mahmood Z, Ahmad S, Mohammad F. Plant-mediated green synthesis of zinc oxide nanoparticles using *Swertia chirayita* leaf extract, characterization and its antibacterial efficacy against some common

pathogenic bacteria. Bionanoscience. 2018;**8**:811-817

[52] Ambika S, Sundrarajan M. Antibacterial behaviour of *Vitex negundo* extract assisted ZnO nanoparticles against pathogenic bacteria. Journal of Photochemistry and Photobiology B-Biology. 2015;**146**:52-57

[53] Anupama C, Kaphle A, Udayabhanu, Nagaraju G. Aegle marmelos assisted facile combustion synthesis of multifunctional ZnO nanoparticles: Study of their photoluminescence, photo catalytic and antimicrobial activities. Journal of Materials Science-Materials in Electronics. 2018;**29**:4238-4249

[54] Bhuyan T, Mishra K, Khanuja M, Prasad R, Varma A. Biosynthesis of zinc oxide nanoparticles from *Azadirachta indica* for antibacterial and photocatalytic applications. Materials Science in Semiconductor Processing. 2015;**32**:55-61

[55] Bayrami A, Parvinroo S, Habibi-Yangjeh A, Pouran SR. Bio-extract-mediated ZnO nanoparticles: Microwave-assisted synthesis, characterization and antidiabetic activity evaluation. Artificial Cells Nanomedicine and Biotechnology. 2018;**46**:730-739

[56] Koupaei MH, Shareghi B, Saboury AA, Davar F, Semnani A, Evini M. Green synthesis of zinc oxide nanoparticles and their effect on the stability and activity of proteinase K. RSC Advances. 2016;**6**:42313-42323

[57] Padalia H, Chanda S. Characterization, antifungal and cytotoxic evaluation of green synthesized zinc oxide nanoparticles using *Ziziphus nummularia* leaf extract. Artificial Cells Nanomedicine and Biotechnology. 2017;**45**:1751-1761

[58] Dobrucka R, Dlugaszewska J, Kaczmarek M. Cytotoxic and antimicrobial effects of biosynthesized ZnO nanoparticles using of *Chelidonium majus* extract. Biomedical Microdevices. 2017;**20**:5

[59] Velmurugan P, Park JH, Lee SM, Yi YJ, Cho M, Jang JS, et al. Eco-friendly approach towards green synthesis of zinc oxide nanocrystals and its potential applications. Artificial Cells Nanomedicine and Biotechnology. 2016;**44**:1537-1543

[60] Elavarasan N, Kokila K, Inbasekar G, Sujatha V. Evaluation of photocatalytic activity, antibacterial and cytotoxic effects of green synthesized ZnO nanoparticles by *Sechium edule* leaf extract. Research on Chemical Intermediates. 2017;**43**:3361-3376

[61] Gupta M, Tomar RS, Kaushik S, Mishra RK, Sharma D. Effective antimicrobial activity of green ZnO nano particles of *Catharanthus roseus*. Frontiers in Microbiology. 2018;**9**:2030

[62] Daphedar A, Taranath TC. Green synthesis of zinc nanoparticles using leaf extract of *Albizia saman* (Jacq.) Merr. and their effect on root meristems of *Drimia indica* (Roxb.) Jessop. Caryologia. 2018;**71**:93-102

[63] Datta A, Patra C, Bharadwaj H, Kaur S, Dimri N, Khajuria R. Green synthesis of zinc oxide nanoparticles using *Parthenium hysterophorus* leaf extract and evaluation of their antibacterial properties. Journal of Biotechnology & Biomaterials. 2017;7:271

[64] Kahraman O, Binzet R, Turunc E, Dogen A, Arslan H. Synthesis, characterization, antimicrobial and electrochemical activities of zinc oxide nanoparticles obtained from *Sarcopoterium spinosum* (L.) spach leaf extract. Materials Research Express. 2018;**5**:115017

[65] Lingaraju K, Naika HR, Manjunath K, Basavaraj RB, Nagabhushana H, Nagaraju G, et al. Biogenic synthesis of zinc oxide nanoparticles using *Ruta graveolens* (L.) and their antibacterial and antioxidant activities. Applied Nanoscience. 2016;**6**:703-710

[66] Moghaddam AB, Moniri M, Azizi S, Rahim RA, Bin Ariff A, Saad WZ, et al. Biosynthesis of ZnO nanoparticles by a new *Pichia kudriavzevii* yeast strain and evaluation of their antimicrobial and antioxidant activities. Molecules. 2017;**22**:872

[67] Mankad M, Patil G, Patel S, Patel D, Patel A. Green synthesis of zinc oxide nanoparticles using *Azadirachta indica* A. Juss. leaves extract and its antibacterial activity against *Xanthomonas orzyae* pv. *oryzae*. Annals of Phytomedicine-an International Journal. 2016;**5**:76-86

[68] Nagajyothi PC, Sreekanth TVM, Tettey CO, Jun YI, Mook SH. Characterization, antibacterial, antioxidant, and cytotoxic activities of ZnO nanoparticles using Coptidis Rhizoma. Bioorganic & Medicinal Chemistry Letters. 2014;**24**:4298-4303

[69] Nagajyothi PC, Cha SJ, Yang IJ, Sreekanth TVM, Kim KJ, Shin HM. Antioxidant and anti-inflammatory activities of zinc oxide nanoparticles synthesized using *Polygala tenuifolia* root extract. Journal of Photochemistry and Photobiology B-Biology. 2015;**146**:10-17

[70] Namvar F, Rahman HS, Rosfarizan Mohamad S, Azizi P, Tahir MSC, et al. Cytotoxic effects of biosynthesized zinc oxide nanoparticles on murine cell lines. Evidence-based Complementary and Alternative Medicine. 2015:593014

[71] Al-Dhabi NA, Arasu MV. Environmentally-friendly green approach for the production of zinc oxide nanoparticles and their anti-fungal, ovicidal, and larvicidal properties. Nanomaterials. 2018;**8**:500

[72] Chung IM, Rahuman AA, Marimuthu S, Kirthi AV, Anbarasan K, Rajakumar G. An investigation of the cytotoxicity and caspase-mediated apoptotic effect of green synthesized zinc oxide nanoparticles using *Eclipta prostrata* on human liver carcinoma cells. Nanomaterials. 2015;**5**:1317-1330

[73] Geetha A, Sakthivel R, Mallika J, Kannusamy R, Rajendran R. Green synthesis of antibacterial zinc oxide nanoparticles using biopolymer *Azadirachta indica* Gum. Oriental Journal of Chemistry. 2016;**32**:955-963

[74] Panda KK, Golari D, Venugopal A, Achary VMM, Phaomei G, Parinandi NL, et al. Green synthesized zinc oxide (ZnO) nanoparticles induce oxidative stress and DNA damage in *Lathyrus sativus* L. root bioassay system. Antioxidants. 2017;**6**:35

[75] Rajakumar G, Thiruvengadam M, Mydhili G, Gomathi T, Chung IM. Green approach for synthesis of zinc oxide nanoparticles from *Andrographis paniculata* leaf extract and evaluation of their antioxidant, anti-diabetic, and anti-inflammatory activities. Bioprocess and Biosystems Engineering. 2018;**41**:21-30

[76] Rehana D, Mahendiran D, Kumar RS, Rahiman AK. In vitro antioxidant and antidiabetic activities of zinc oxide nanoparticles synthesized using different plant extracts. Bioprocess and Biosystems Engineering. 2017;**40**:943-957

[77] Sharmila G, Muthukumaran C, Sandiya K, Santhiya S, Pradeep RS, Kumar NM, et al. Biosynthesis, characterization, and antibacterial activity of zinc oxide nanoparticles derived from *Bauhinia tomentosa* leaf extract. Journal of Nanostructure in Chemistry. 2018;**8**:293-299

[78] Shoeb M, Singh BR, Khan JA, Khan W, Singh BN, Singh HB, et al. ROS-dependent anticandidal activity of zinc oxide nanoparticles synthesized by using egg albumen as a biotemplate. Advances in Natural Sciences-Nanoscience and Nanotechnology. 2013;**4**:035015

[79] Surendra TV, Roopan SM, Al-Dhabi NA, Arasu MV, Sarkar G, Suthindhiran K. Vegetable peel waste for the production of ZnO nanoparticles and its toxicological efficiency, antifungal, hemolytic, and antibacterial activities. Nanoscale Research Letters. 2016;**11**:546

[80] Suresh D, Nethravathi PC, Udayabhanu, Rajanaika H, Nagabhushana H, Sharma SC. Green synthesis of multifunctional zinc oxide (ZnO) nanoparticles using *Cassia fistula* plant extract and their photodegradative, antioxidant and antibacterial activities. Materials Science in Semiconductor Processing. 2015;**31**:446-454

[81] Nava OJ, Luque PA, Gomez-Gutierrez CM, Vilchis-Nestor AR, Castro-Beltran A, Mota-Gonzalez ML, et al. Influence of *Camellia sinensis* extract on zinc oxide nanoparticle green synthesis. Journal of Molecular Structure. 2017;**1134**:121-125

[82] Paul B, Vadivel S, Dhar SS, Debbarma S, Kumaravel M. One-pot green synthesis of zinc oxide nano rice and its application as sonocatalyst for degradation of organic dye and synthesis of 2-benzimidazole derivatives. Journal of Physics and Chemistry of Solids. 2017;**104**:152-159

[83] Prasad AR, Ammal PR, Joseph A. Effective photocatalytic removal of different dye stuffs using green synthesized zinc oxide nanogranules. Materials Research Bulletin. 2018;**102**:116-121

[84] Vidya C, Manjunatha C, Chandraprabha MN, Rajshekar M, Raj M. Hazard free green synthesis of ZnO nano-photo-catalyst using *Artocarpus heterophyllus* leaf extract for the degradation of Congo red dye in water treatment applications. Journal of Environmental Chemical Engineering. 2017;**5**:3172-3180

[85] Hassan SSM, El Azab WIM, Ali HR, Mansour MSM. Green synthesis and characterization of ZnO nanoparticles for photocatalytic degradation of anthracene. Advances in Natural Sciences-Nanoscience and Nanotechnology. 2015;**6**:045012

[86] Nava OJ, Soto-Robles CA, Gomez-Gutierrez CM, Vilchis-Nestor AR, Castro-Beltran A, Olivas A, et al. Fruit peel extract mediated green synthesis of zinc oxide nanoparticles. Journal of Molecular Structure. 2017;**1147**:1-6

[87] Ngoepe NM, Mbita Z, Mathipa M, Mketo N, Ntsendwana B, Hintsho-Mbita NC. Biogenic synthesis of ZnO nanoparticles using *Monsonia burkeana* for use in photocatalytic, antibacterial and anticancer applications. Ceramics International. 2018;**44**:16999-17006

[88] Ishwarya R, Vaseeharan B, Kalyani S, Banumathi B, Govindarajan M, Alharbi NS, et al. Facile green synthesis of zinc oxide nanoparticles using *Ulva lactuca* seaweed extract and evaluation of their photocatalytic, antibiofilm and insecticidal activity. Journal of Photochemistry and Photobiology B-Biology. 2018;**178**:249-258

[89] Ali J, Irshad R, Li BS, Tahir K, Ahmad A, Shakeel M, et al. Synthesis and characterization of phytochemical fabricated zinc oxide nanoparticles with enhanced antibacterial and catalytic applications. Journal of Photochemistry and Photobiology B-Biology. 2018;**183**:349-356

[90] Stan M, Popa A, Toloman D, Dehelean A, Lung I, Katona G. Enhanced photocatalytic degradation

properties of zinc oxide nanoparticles synthesized by using plant extracts. Materials Science in Semiconductor Processing. 2015;**39**:23-29

[91] Aminuzzaman M, Ying LP, Goh WS, Watanabe A. Green synthesis of zinc oxide nanoparticles using aqueous extract of Garcinia mangostana fruit pericarp and their photocatalytic activity. Bulletin of Materials Science. 2018;**41**:50

[92] Fu L, Fu ZX. Plectranthus amboinicus leaf extract-assisted biosynthesis of ZnO nanoparticles and their photocatalytic activity. Ceramics International. 2015;**41**:2492-2496

[93] Gawade VV, Gavade NL, Shinde HM, Babar SB, Kadam AN, Garadkar KM. Green synthesis of ZnO nanoparticles by using *Calotropis procera* leaves for the photodegradation of methyl orange. Journal of Materials Science-Materials in Electronics. 2017;**28**:14033-14039

[94] Kumar B, Smita K, Cumbal L, Debut A. Green approach for fabrication and applications of zinc oxide nanoparticles. Bioinorganic Chemistry and Applications. 2014;**2014**:523869

[95] Phukan S, Bomjen P, Shripathi T, Rashid MH. Green route biosynthesis of shape-tunable ZnO nanostructures and their photocatalytic applications. ChemistrySelect. 2017;**2**:11137-11147

[96] Rao MD, Gautam P. Synthesis and characterization of ZnO nanoflowers using *Chlamydomonas reinhardtii*: A green approach. Environmental Progress & Sustainable Energy. 2016;**35**:1020-1026

[97] Soto-Robles CA, Nava OJ, Vilchis-Nestor AR, Castro-Beltran A, Gomez-Gutierrez CM, Lugo-Medina E, et al. Biosynthesized zinc oxide using *Lycopersicon esculentum* peel extract for methylene blue degradation. Journal of Materials Science-Materials in Electronics. 2018;**29**:3722-3729

[98] Zheng YH, Fu L, Han FG, Wang AW, Cai W, Yu JP, et al. Green biosynthesis and characterization of zinc oxide nanoparticles using *Corymbia citriodora* leaf extract and their photocatalytic activity. Green Chemistry Letters and Reviews. 2015;**8**:59-63

[99] Kalaiselvi A, Roopan SM, Madhumitha G, Ramalingam C, Al-Dhabi NA, Arasu MV. Catharanthus roseus-mediated zinc oxide nanoparticles against photocatalytic application of phenol red under UV@ 365 nm. Current Science. 2016;**111**:1811-1815

[100] Azizi S, Shahri MM, Mohamad R. Green Synthesis of zinc oxide nanoparticles for enhanced adsorption of lead ions from aqueous solutions: Equilibrium, kinetic and thermodynamic studies. Molecules. 2017;**22**:831

[101] Fazlzadeh M, Khosravi R, Zarei A. Green synthesis of zinc oxide nanoparticles using *Peganum harmala* seed extract, and loaded on *Peganum harmala* seed powdered activated carbon as new adsorbent for removal of Cr(VI) from aqueous solution. Ecological Engineering. 2017;**103**:180-190

[102] Muthuchamy N, Atchudan R, Edison T, Perumal S, Lee YR. High-performance glucose biosensor based on green synthesized zinc oxide nanoparticle embedded nitrogen-doped carbon sheet. Journal of Electroanalytical Chemistry. 2018;**816**:195-204

[103] Sharma D, Sabela MI, Kanchi S, Bisetty K, Skelton AA, Honarparvar B. Green synthesis, characterization and electrochemical sensing of silymarin by ZnO nanoparticles:

Experimental and DFT studies. Journal of Electroanalytical Chemistry. 2018;**808**:160-172

[104] Jayaseelan C, Rahuman AA, Kirthi AV, Marimuthu S, Santhoshkumar T, Bagavan A, et al. Novel microbial route to synthesize ZnO nanoparticles using *Aeromonas hydrophila* and their activity against pathogenic bacteria and fungi. Spectrochimica Acta Part A-Molecular and Biomolecular Spectroscopy. 2012;**90**:78-84

[105] Singh BN, Rawat AKS, Khan W, Naqvi AH, Singh BR. Biosynthesis of stable antioxidant ZnO nanoparticles by *Pseudomonas aeruginosa* rhamnolipids. PLoS One. 2014;**9**:e106937

[106] Kundu D, Hazra C, Chatterjee A, Chaudhari A, Mishra S. Extracellular biosynthesis of zinc oxide nanoparticles using *Rhodococcus pyridinivorans* NT2: Multifunctional textile finishing, biosafety evaluation and in vitro drug delivery in colon carcinoma. Journal of Photochemistry and Photobiology B-Biology. 2014;**140**:194-204

[107] Tripathi RM, Bhadwal AS, Gupta RK, Singh P, Shrivastav A, Shrivastav BR. ZnO nanoflowers: Novel biogenic synthesis and enhanced photocatalytic activity. Journal of Photochemistry and Photobiology B-Biology. 2014;**141**:288-295

[108] Agarwal H, Kumar SV, Rajeshkumar S. A review on green synthesis of zinc oxide nanoparticles—An eco-friendly approach. Resource-Efficient Technologies. 2017;**3**:406-413

[109] Nagarajan S, Kuppusamy KA. Extracellular synthesis of zinc oxide nanoparticle using seaweeds of gulf of Mannar, India. Journal of Nanobiotechnology. 2013;**11**:39

[110] Raliya R, Tarafdar JC. ZnO nanoparticle biosynthesis and its effect on phosphorous-mobilizing enzyme secretion and gum contents in clusterbean (*Cyamopsis tetragonoloba* L.). Agricultural Research. 2013;**2**:48-57

[111] Dhandapani P, Siddarth AS, Kamalasekaran S, Maruthamuthu S, Rajagopal G. Bio-approach: Ureolytic bacteria mediated synthesis of ZnO nanocrystals on cotton fabric and evaluation of their antibacterial properties. Carbohydrate Polymers. 2014;**103**:448-455

[112] Azizi S, Ahmad MB, Namvar F, Mohamad R. Green biosynthesis and characterization of zinc oxide nanoparticles using brown marine macroalga Sargassum muticum aqueous extract. Materials Letters. 2014;**116**:275-277

[113] Gharagozlou M, Baradaran Z, Bayati R. A green chemical method for synthesis of ZnO nanoparticles from solid-state decomposition of Schiff-bases derived from amino acid alanine complexes. Ceramics International. 2015;**41**:8382-8387

[114] Alnarabiji MS, Yahya N, Hamed Y, Ardakani SEM, Azizi K, Klemes JJ, et al. Scalable bio-friendly method for production of homogeneous metal oxide nanoparticles using green bovine skin gelatin. Journal of Cleaner Production. 2017;**162**:186-194

[115] Fardood ST, Ramazani A, Moradi S, Asiabi PA. Green synthesis of zinc oxide nanoparticles using arabic gum and photocatalytic degradation of direct blue 129 dye under visible light. Journal of Materials Science-Materials in Electronics. 2017;**28**:13596-13601

[116] Darroudi M, Sabouri Z, Oskuee RK, Zak AK, Kargar H, Hamid MHNA. Sol-gel synthesis, characterization, and neurotoxicity effect of zinc oxide nanoparticles using gum tragacanth. Ceramics International. 2013;**39**:9195-9199

[117] Somu P, Paul S. Casein based biogenic-synthesized zinc oxide nanoparticles simultaneously decontaminate heavy metals, dyes, and pathogenic microbes: A rational strategy for wastewater treatment. Journal of Chemical Technology and Biotechnology. 2018;**93**:2962-2976

[118] Subramanian N, Al Ghaferi A. A green synthetic route for zinc oxide nanoarchitectures using L-lysine. Materials Letters. 2013;**92**:361-364

[119] Xu J, Su H, Han J, Chen Y, Song W, Gu Y, et al. In situ deposition of flower-like ZnO on silk fibroin fibers. Applied Physics A: Materials Science and Processing. 2012;**108**:235-238

[120] Vijayakumar S, Vaseeharan B. Antibiofilm, anti cancer and ecotoxicity properties of collagen based ZnO nanoparticles. Advanced Powder Technology. 2018;**29**:2331-2345

[121] Arfat YA, Benjakul S, Prodpran T, Sumpavapol P, Songtipya P. Properties and antimicrobial activity of fish protein isolate/fish skin gelatin film containing basil leaf essential oil and zinc oxide nanoparticles. Food Hydrocolloids. 2014;**41**:265-273

[122] Chan ES, Lee BB, Ravindra P, Poncelet D. Prediction models for shape and size of Ca-alginate macrobeads produced through extrusion-dripping method. Journal of Colloid and Interface Science. 2009;**338**:63-72

[123] Goes MF, Sinhoreti MA, Consani S, Silva MA. Morphological effect of the type, concentration and etching time of acid solutions on enamel and dentin surfaces. Brazilian Dental Journal. 1998;**9**:3-10

[124] Leone F, Gignone A, Ronchetti S, Cavalli R, Manna L, Banchero M, et al. A green organic-solvent-free route to prepare nanostructured zinc oxide carriers of clotrimazole for pharmaceutical applications. Journal of Cleaner Production. 2018;**172**:1433-1439

[125] Kolodziejczak-Radzimska A, Jesionowski T. Zinc oxide-from synthesis to application: A review. Materials. 2014;**7**:2833-2881

[126] Kakiuchi K, Hosono E, Kimura T, Imai H, Fujihara S. Fabrication of mesoporous ZnO nanosheets from precursor templates grown in aqueous solutions. Journal of Sol-Gel Science and Technology. 2006;**39**:63-72

[127] Horzum N, Hilal ME, Isik T. Enhanced bactericidal and photocatalytic activities of ZnO nanostructures by changing the cooling route. New Journal of Chemistry. 2018;**42**:11831-11838

[128] Sabbaghan M, Sadat A. Green synthesis of symmetrical imidazolium based ionic liquids and their application in the preparation of ZnO nanostructures. Ceramics International. 2016;**42**:3820-3825

[129] Sundrarajan M, Jegatheeswaran S, Selvam S, Gowri R, Balaji M, Bharathi K. Green approach: Ionic liquid assisted synthesis of nanocrystalline ZnO in phyto medium and their antibacterial investigation. Materials Letters. 2017;**201**:31-34

[130] Duan X, Ma J, Zheng W. The art of using ionic liquids in the synthesis of inorganic nanomaterials. CrystEngComm. 2014;**16**:2550-2559

[131] Li Q Kumar V, Li Y, Zhang H, Marks TJ, Chang RPH. Fabrication of ZnO nanorods and nanotubes in aqueous solutions. Chemistry of Materials. 2005;**17**:1001-1006

[132] Huddleston JG, Willauer HD, Swatloski RP, Visser AE, Rogers RD. Room temperature ionic liquids

as novel media for 'clean' liquid–liquid extraction. Chemical Communications. 1998;**0**:1765-1766

[133] Sabbaghan M, Sadat A, Elahe S. Synthesis and optical properties of ZnO nanostructures in imidazolium-based ionic liquids. Solid State Sciences. 2012;**14**:1191-1195

[134] Alammar T, Mudring A-V. Sonochemical synthesis of 0D, 1D, and 2D zinc oxide nanostructures in ionic liquids and their photocatalytic activity. ChemSusChem. 2011;**16**:1796-1804

[135] Bohle DS, Spina CJ. Chelating the surface of zinc in zinc oxide nanocrystals: Spectroscopic characterization of ZnO surface-bound eriochrome black T and 8-hydroxyquinoline. The Journal of Physical Chemistry C. 2009;**72**:14435-14439

[136] Eadi SB, Kim S, Jeong SW. Effect of surfactant on growth of ZnO nanodumbbells and their characterization. Journal of Chemistry. 2017;**2017**:1-8

[137] Amde M, Tan Z-q, Liu R, Liu J-f. Nanofluid of zinc oxide nanoparticles in ionic liquid for single drop liquid microextraction of fungicides in environmental waters prior to high performance liquid chromatographic analysis. Journal of Chromatography A. 2015;**1395**:7-15

[138] Spietelun A, Marcinkowski Ł, De La Guardia M, Namieśnik J. Green aspects, developments and perspectives of liquid phase microextraction techniques. Talanta. 2014;**119**:34-45

[139] Zhang CL, Zhou WN, Hang Y, Lü Z, Hou HD, Zuo YB, et al. Hydrothermal growth and characterization of ZnO crystals. Journal of Crystal Growth. 2008;**310**:1819-1822

[140] Guo TH, Liu Y, Zhang YC, Zhang M. Green hydrothermal synthesis and optical absorption properties of ZnO_2 nanocrystals and ZnO nanorods. Materials Letters. 2011;**65**:639-641

[141] Al-Sabahi J, Bora T, Al-Abri M, Dutta J. Controlled defects of zinc oxide nanorods for efficient visible light photocatalytic degradation of phenol. Materials. 2016;**9**:1-10

[142] Baruah S, Dutta J. Hydrothermal growth of ZnO nanostructures. Science and Technology of Advanced Materials. 2009;**10**:013001

[143] Baruah S, Mahmood MA, Myint MTZ, Bora T, Dutta J. Enhanced visible light photocatalysis through fast crystallization of zinc oxide nanorods. Beilstein Journal of Nanotechnology. 2010;**1**:14-20

[144] Al-Sabahi J, Bora T, Claereboudt M, Al-Abri M, Dutta J. Visible light photocatalytic degradation of HPAM polymer in oil produced water using supported zinc oxide nanorods. Chemical Engineering Journal. 2018;**351**:56-64

[145] Singh S, Srivastava VC, Lo SL, Mandal TK, Naresh G. Morphology-controlled green approach for synthesizing the hierarchical self-assembled 3D porous ZnO superstructure with excellent catalytic activity. Microporous and Mesoporous Materials. 2017;**239**:296-309

[146] Lam S-m, Sin J-c. A green and facile hydrothermal synthesis of ZnO nanorods for photocatalytic application. JOJ Material Science. 2018;**4**:8-10

[147] Straumal B, Mazilkin A, Protasova S, Myatiev A, Straumal P, Goering E, et al. Influence of texture on the ferromagnetic properties of nanograined ZnO films. Physica Status Solidi b. 2011;**248**:1581-1586

[148] Lam SM, Sin JC, Zuhairi Abdullah A, Rahman Mohamed A.

Green hydrothermal synthesis of ZnO nanotubes for photocatalytic degradation of methylparaben. Materials Letters. 2013;**93**:423-426

[149] Kumar V, Gohain M, Som S, Kumar V, Bezuindenhoudt BCB, Swart HC. Microwave assisted synthesis of ZnO nanoparticles for lighting and dye removal application. Physica B: Condensed Matter. 2016;**480**:36-41

[150] Polshettiwar V, Nadagouda MN, Varma RS. Microwave-assisted chemistry: A rapid and sustainable route to synthesis of organics and nanomaterials. Australian Journal of Chemistry. 2009;**62**:16-26

[151] Azizi S, Mohamad R, Shahri MM, McPhee DJ. Green microwave-assisted combustion synthesis of zinc oxide nanoparticles with *Citrullus colocynthis* (L.) schrad: Characterization and biomedical applications. Molecules. 2017;**22**:1-13

[152] Sutradhar P, Debbarma M, Saha M. Microwave synthesis of zinc oxide nanoparticles using coffee powder extract and its application for solar cell. Synthesis and Reactivity in Inorganic, Metal-Organic and Nano-Metal Chemistry. 2016;**46**:1622-1627

[153] Iwamura T, Goto S-i, Sakaguchi M, Chujo Y. Synthesis of submicrometer zinc oxide particles and zinc oxide nanowires using microwave irradiation. Chemistry Letters. 2016;**45**:508-510

[154] Salari Z, Ameri A, Forootanfar H, Adeli-Sardou M, Jafari M, Mehrabani M, et al. Microwave-assisted biosynthesis of zinc nanoparticles and their cytotoxic and antioxidant activity. Journal of Trace Elements in Medicine and Biology. 2017;**39**:116-123

[155] Tang J, Chai J, Huang J, Deng L, Nguyen XS, Sun L, et al. ZnO nanorods with low intrinsic defects and high optical performance grown by facile microwave-assisted solution method. ACS Applied Materials and Interfaces. 2015;**7**:4737-4743

[156] Flint EB, Suslick KS. The temperature of cavitation. Science. 1991;**253**:1397-1399

[157] Oxley JD, Prozorov T, Suslick KS. Sonochemistry and sonoluminescence of room-temperature ionic liquids. Journal of the American Chemical Society. 2003;**125**:11138-11139

[158] Zak AK, Majid WH, Wang HZ, Yousefi R, Golsheikh AM, Ren ZF. Sonochemical synthesis of hierarchical ZnO nanostructures. Ultrasonics Sonochemistry. 2013;**20**:395-400

[159] Bhatte KD, Sawant DN, Pinjari DV, Pandit AB, Bhanage BM. One pot green synthesis of nano sized zinc oxide by sonochemical method. Materials Letters. 2012;**77**:93-95

[160] Zhang YC, Wu X, Hu XY, Guo R. Low-temperature synthesis of nanocrystalline ZnO by thermal decomposition of a green single-source inorganic precursor in air. Journal of Crystal Growth. 2005;**280**:250-254

[161] Srikant V, Clarke DR. On the optical band gap of zinc oxide. Journal of Applied Physics. 1998;**83**:5447-5451

[162] Marinho JZ, Romeiro FC, Lemos SCS, Motta FV, Riccardi CS, Li MS, et al. Urea-based synthesis of zinc oxide nanostructures at low temperature. Journal of Nanomaterials. 2012;**2012**:427172

[163] Raja M, Shanmugaraj AM, Ryu SH. Preparation of template free zinc oxide nanoparticles using sol-gel chemistry. Journal of Nanoscience and Nanotechnology. 2008;**8**:4224-4226

[164] Yan JK, Wang YY, Zhu L, Wu JY. Green synthesis and characterization of zinc oxide nanoparticles using carboxylic curdlan and their interaction with bovine serum albumin. RSC Advances. 2016;**6**:77752-77759

Section 3

ZnO Doping

Chapter 4

Doped Zinc Oxide Nanostructures for Photovoltaic Solar Cells Application

Tyona MD

Abstract

Zinc oxide and doping effects of Cu on its structural, morphological, optical, and surface wettability properties and the consequent influence on photoelectrochemical solar cell performance has been reviewed. Cu dopant in the doping solution is varied in the range of 1 to 5 at.% which significantly affected the properties of ZnO. Slight changes in the lattice parameters of the Cu-doped zinc oxide (CZO) electrodes were reported, due to the successful substitution of Zn^{2+} by Cu^{2+} and also enhancement in crystallinity of the films at 3 at.% Cu due to reduction in crystallographic defects in the film. Surface morphologies were reported with densely grown nanorods over the varied range of Cu, with 3 at.% having the densest microstructures with average diameter approximately 125 nm. A review of optical properties indicated significant enhancement in absorption edge of approximately 60 nm into the visible band for the nanorods with 3 at.% Cu content due to light scattering. Optical energy band-gaps decrease from 3.03 to 2.70 eV with Cu doping. Surface wettability was adjudged hydrophilic for all the films, implying high porosity and water contact angles depended on Cu content. Photoelectrochemical cell performance indicated an n-type photoactivity in sodium sulfate (Na_2SO_4) electrolyte, which motivates to check its feasibility in solar cell applications.

Keywords: zinc oxide, nanostructures, CZO, photoelectrochemical solar cells, Cu concentration, nanorods

1. Introduction

Zinc oxide is an inorganic compound having a chemical formula ZnO. It is a white powder which is nearly insoluble in water. It crystallizes in two main forms, the hexagonal wurtzite and cubic zinc blende. The wurtzite structure with lattice parameters a = 0.3296 and c = 0.52065 nm is found to be more stable than the zinc blende structure, and hence it is more widely used [1]. The ZnO structure is commonly described as consisting of a number of alternating planes composed of tetrahedrally coordinated O^{2-} and Zn^{2+} ions, stacked alternately along the c-axis without a central symmetry as illustrated in **Figure 1** [1, 2]. It is a group II–VI semiconductor with a wide band gap of about 3.33 eV. Due to its direct and wide band gap in the near-UV spectral region [3–5] and a large free exciton binding energy, it has become a promising functional semiconductor material, which possesses a wide range of novel applications. ZnO has been identified with many unique properties

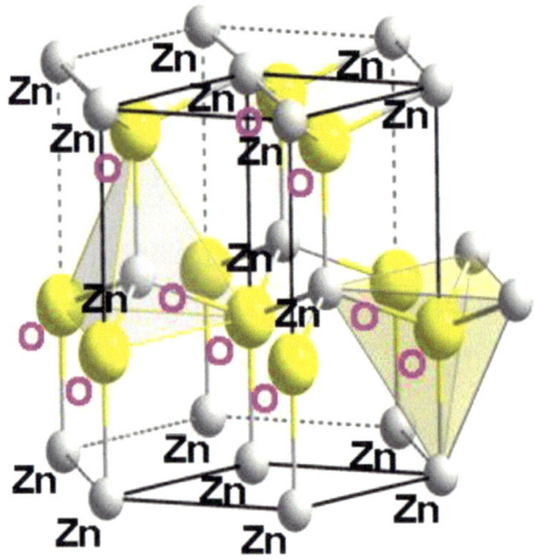

Figure 1.
Hexagonal Wurtzite crystal structure of ZnO [1].

such as excitonic emission at or even above room temperature, optical transparency in the visible range, high surface-to-volume ratio and quantum confinement effect [6], amongst others, which have motivated intensive study of the semiconductor during the last two decades. ZnO is mostly known to crystalize as an n-type semiconductor, whereas synthesis of the p-type is not generally easy [1, 7].

ZnO is simple to synthesize; both chemical and physical techniques are used to produce excellent epitaxial films. The most commonly used techniques to grow epitaxial films of ZnO include electrodeposition, spray pyrolysis, sol–gel process, successive ionic layer adsorption and reaction (SILAR), RF sputtering, chemical bath deposition (CBD), spin coating, electron beam epitaxy, laser evaporation and ion beam sputtering, amongst others [7, 8]. **Figure 2** illustrates the various synthetic techniques (chemical as well as physical) that are generally used to grow compound and alloys of ZnO. The choice of a particular technique would be guided by some factors such as the application intended for the synthesis, effectiveness of the technique and cost implication [10, 11]. ZnO has been identified as one of the semiconductors with the largest number of novel nanostructures such as nanocombs, nanorings, nanohelixes/nanosprings, nanobelts, nanowires, nanorods, nanotubes, nanocages, etc., with a wide range of technological applications [12–15]. Novel applications of ZnO nanostructures include optical modulator waveguide, photonic crystals, surface acoustic wave filters, varistors, photodetectors, gas sensors, light-emitting diode, photodiodes and solar cells, amongst others [12].

Photovoltaic (PV) application of ZnO nanostructures requires large internal surface area with porous and high surface roughness to support good penetration of electrolyte [13, 14]. Chemical techniques are very simple, much reliable and cost-effective for the synthesis of high-quality electrodes for PV application. Most especially, chemical bath deposition technique is very suitable for growing large area films of ZnO with fascinating properties for photoelectrochemical solar cells [15, 16]. This technique is suitable for growing ZnO nanostructures on many substrates including microscope glass and stainless steel [6].

In several applications such as optoelectronics, ZnO can be used as a complement or alternative to some semiconductors such as GaN, and many researches are ongoing globally to further improve the properties of the semiconductor [10]. Trying to control the unintentional n-type conductivity and to achieve p-type

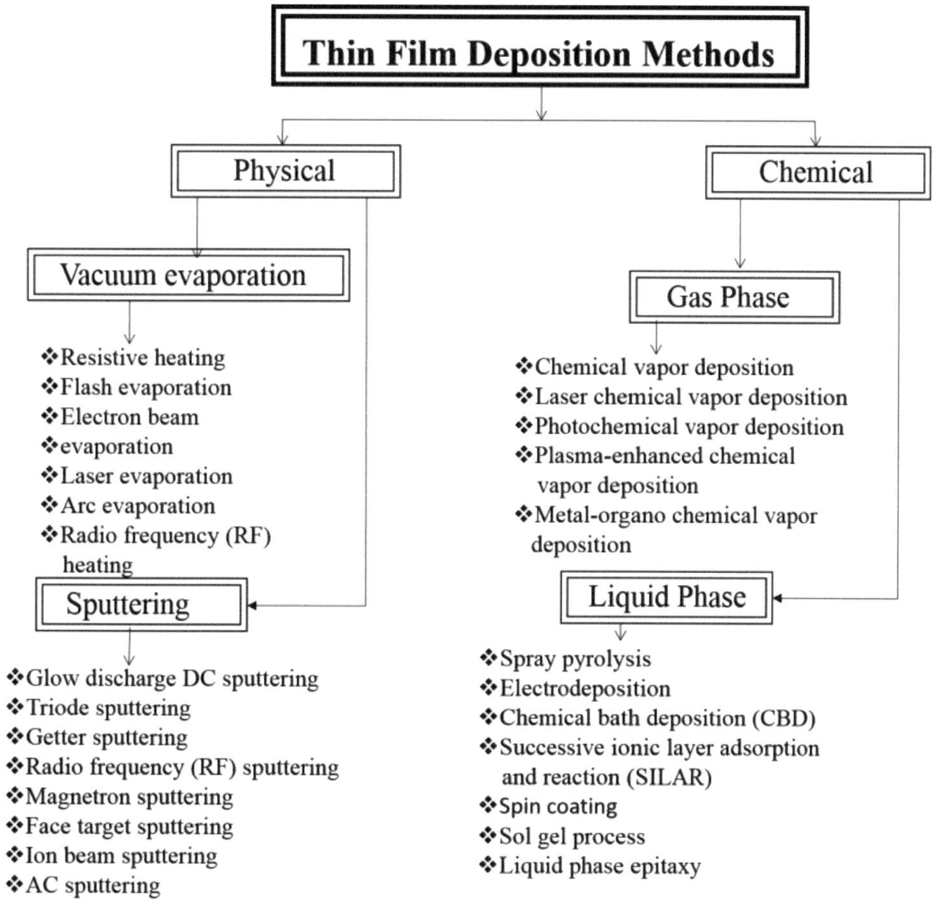

Figure 2.
Broad classification of thin film deposition techniques [9].

conductivity are such famous research themes. Other approaches such as first-principles calculations based on density functional theory (DFT) are theoretical and, however, are useful to provide in-depth understanding of the role of native point defects and impurities on the unintentional n-type conductivity in ZnO [14, 17]. Acceptor doping in ZnO which will lead to stable p-type is not yet well known [18].

It has been noted that wide application of ZnO in electronic devices has been limited by the lack of inadequate control over its electrical conductivity [10, 19]. Controlling the conductivity in ZnO can be achieved by means of band-gap engineering [10]. Introducing small concentrations of native point defects and impurities (dopants) (down to 10–14 cm^{-3} or 0.01 ppm) can significantly affect the electrical, structural, optical and morphological properties of the semiconductors [14, 18]. Therefore, understanding the role of native point defects (i.e. vacancies, interstitials and antisites) and the incorporation of impurities (doping) is the key towards controlling the conductivity in ZnO, which in effect alters the band gap, thus enhancing its performance [1].

Band-gap engineering of ZnO can also be achieved by alloying with MgO or CdO. The band gap of ZnO is increased with the addition of Mg, whereas the addition of Cd decreases the band gap, which is similar to the effects of Al and In in GaN [1, 18]. It is well known that MgO and CdO crystallize in the rock salt structure;

however, alloys of $Mg_{1-x}Zn_xO$ and $Cd_{1-x}Zn_xO$ with moderate concentrations will assume the wurtzite structure of the parent compound with significant band-gap variation [1]. This chapter will carry out a detailed review of the doping effects of Cu, Al and In on ZnO and the influence of the doped electrodes on the PEC solar cell performance.

2. Doping as a technique for engineering structural, optical and morphological properties of ZnO

Doping implies the deliberate inclusion of impurities into the crystal structure of a semiconductor in order to improve its conductivity and modify some of its characteristics [19]. For elemental semiconductors such as silicon and germanium, the commonly used dopants include boron, aluminum and indium (trivalent elements) and phosphorus, arsenic and antimony (pentavalent element) [20]. In the process of doping, the dopant is integrated into the lattice structure of the semiconductor crystal. The number of valence electrons of the dopant defines the type of doping that would be achieved [20]. Doping a semiconductor with a trivalent element results into p-type doping, whereas using a pentavalent element produces an n-type doping as illustrated in **Figure 3**. For an n-doping, electrons are the majority charge carriers, while holes are the majority carriers in p-doping. The conductivity of a silicon crystal which is properly doped can be increased by a factor of 10^6 [1].

Compound semiconductors such as ZnO can also be doped with the same or similar dopants like copper and indium. There are reports in the literature on the modification of structural, morphological and optical properties of ZnO by doping with Al, Cu or In. Cu and In dopants have been confirmed to lower the band gap of ZnO appreciably [10, 18].

Doped semiconductors are electrically neutral. The terms n- and p-type doped do only refer to the majority charge carriers. Each positive or negative charge carrier belongs to a fixed negative or positive charged dopant as illustrated in **Figure 4**.

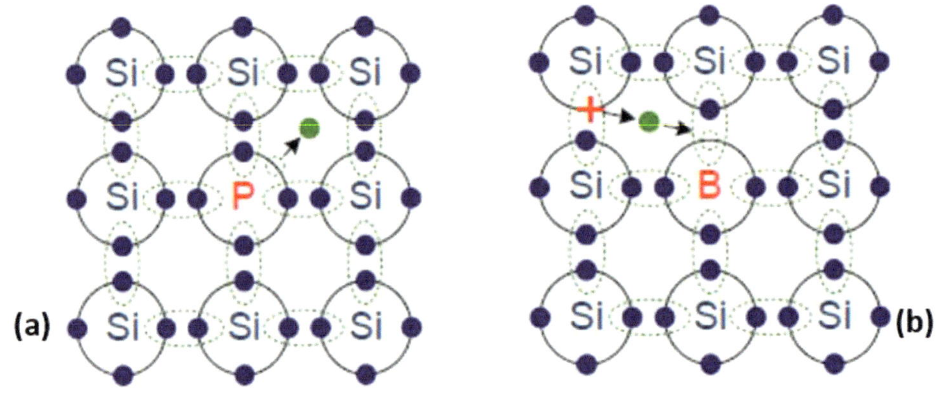

Figure 3.
(a) N-doping with phosphorus and (b) p-doping with boron [2].

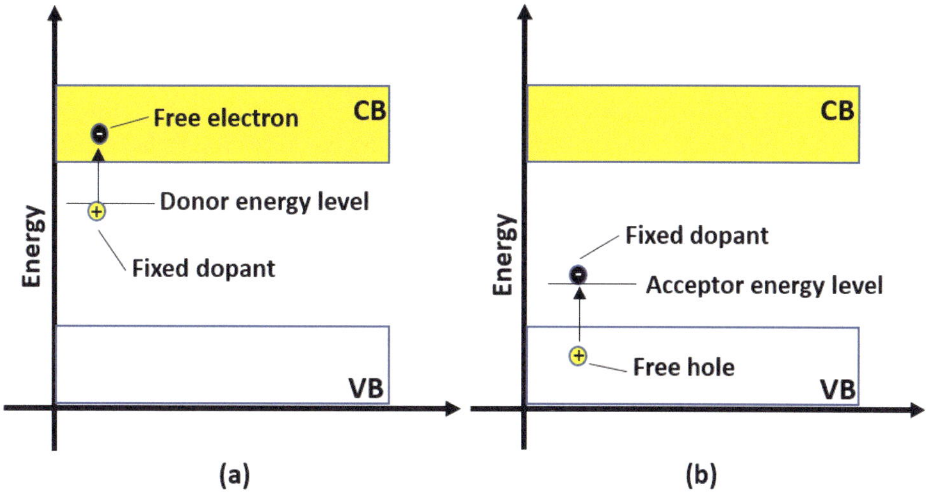

Figure 4.
Doped semiconductor showing energy levels of (a) n-type doping (b) p-type doping [20].

3. Effects of Cu dopant on ZnO and the PEC solar cell performance of Cu-doped ZnO (CZO) electrodes

Wide band-gap semiconductors such as ZnO and TiO_2 (3.3 eV) are suitable for many semiconductor applications such as PEC solar cells due to their thermal, photo- and electrochemical stability and resistance against atmospheric corrosion [21]. However, the wide band gap in such semiconductors is a drawback on their light absorption capability because only photons below a threshold wavelength λg can be absorbed, since the solar spectrum has its maximum intensity at about 2.7 eV [19]. Previous investigations have confirmed that band gap in ZnO semiconductor can be controlled by doping with appropriate dopants [10]. This can also modify optical and structural properties of the semiconductor to meet pre-desired applications [19].

Transitional metals are good dopants; however, Cu and Al are prevalently studied as dopants for ZnO [18]. Cu is a highly conducting metal with conductivity higher than that of Al and can enhance green luminescence band through creation of localized states in the band gap of ZnO. It is also known that due to the high ionization energy and low formation energy of Cu, it can rapidly substitute Zn in ZnO lattice [22].

Tyona et al. [10] investigated the effect of Cu doping on optoelectronic properties of chemically synthesized ZnO electrodes. These properties of Cu-doped ZnO nanostructures were influenced by various parameters such as growth conditions, Cu concentration and post-growth annealing. Cu concentration in ZnO was varied in the range of 1–5%. This quantity may be small; however, it produces significant physical changes in ZnO, and it is considered to be within a strict doping range of up to 10%. Beyond this range, such a reaction may be turning towards composite growth or alloys [10].

Their experimental procedures showed that $Zn(NO_3)_2.6H_2O$ (SD Fine Chemicals) was used as the source of Zn^{2+} and $CuCl_2.2H_2O$ (Chemco Fine, India) as the source of Cu^{2+}, and NH_3 solution (28%) (Thomas Baker) was the complexing agent. An aqueous solution of 0.1 M $Zn(NO_3)_2.6H_2O$ was prepared, and cupric chloride dihydrate ($CuCl_2.2H_2O$) was added. Aqueous NH_3 solution (28%) was used as the complexing agent. The solution was maintained at a pH $\approx$ 11.5. Microscopic glass slides and stainless steel slides were used as substrates and immersed vertically in the solution using Bakelite holder at a bath temperature of 353 K. The substrates were coated with CZO thin films, well adherent to the substrates after 5 h, washed, dried in air and preserved in a vacuum desiccator. Further, as-deposited films were air annealed at 673 K for 2 h and characterized using the following techniques: X-ray diffraction (XRD) patterns were obtained with Cu Kα (λ = 1.5406 Å) radiation from a Philips X-ray diffractometer, Philips PW1830, in the range 20–80°. The morphology of the CZO thin films was measured with the scanning electron microscope (SEM) using JEOL JSM-6360. Optical properties were studied using Shimadzu UV-1800 spectrophotometer in the range 300–800 nm. Contact angle meter (ramé-hart USA equipment) with CCD camera was used to measure the surface wettability of the films. PEC activities of CZO films were studied by forming cells with n-CZO (stainless steel substrate)/0.1 M Na_2SO_4/platinum/SCE, which were illuminated with an 80 mW/cm^2 xenon arc lamp [10]. They analyzed their results under the following headings:

3.1 Film formation mechanism

In their results, they examined CZO film formation process by chemical bath deposition method under four steps of particle growth [10] as in **Figure 5**: (a) nucleation, (b) aggregation, (c) coalescence and, subsequently, (d) growth by stacking of the particles (**Figure 5e**). It is noted that during growth process, the diameter and density of ZnO nanorod are highly affected by the density of the nucleation sites and the pH value of the aqueous solution. Therefore, introducing Cu impurities into the reaction path would increase the nucleation density and hence enhance the growth rate, which, in turn, results in a coarsening and lateral aggregation of the nanorods [10]. This is apparently due to the fact that Cu^{2+}

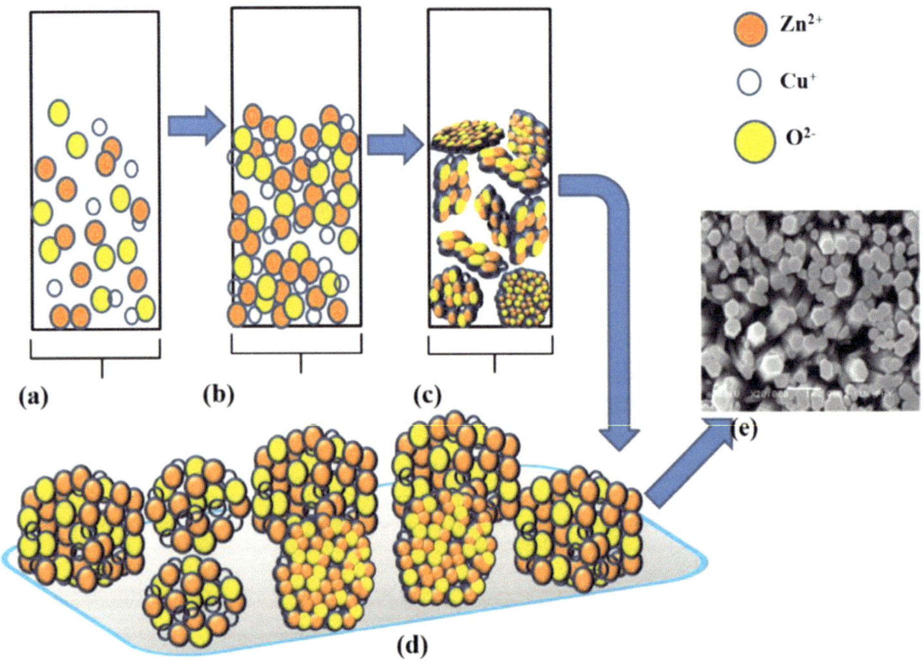

Figure 5.
The schematic growth model for the formation of CZO thin films: (a) nucleation, (b) aggregation, (c) coalescence, (d) subsequent growth by stacking of the particles and (e) surface view of grown stacked nanostructures [10].

increases the number of free metal ions in the solution which forms nucleation sites and the highly alkaline medium helps to speed up the release of chalcogenide ions in the solution which cause the increase in film growth rate. During nucleation, the heterogeneous reaction at the substrate surface takes place when molecule clusters begin to undergo rapid decomposition and particles combine to grow up to a certain thickness of the film [10]. For deposition of ZnO, $Zn(NO_3)_2.6H_2O$ was used as a source of Zn^{2+} ions. When ammonia was added to it, white precipitate of $Zn(OH)_2$ occurred, and further $Zn(OH)_2$ redissolved into the solution by addition of excess ammonia which results into formation of zinc tetra amino complex ($[Zn(NH_3)4]^{2+}$). The decomposition of $[Zn(NH_3)_4]^{2+}$ releases Zn^{2+} ions, which reacts with OH^- ions in the solution and results in the formation of $Zn(OH)_2$ or ZnO particles. The possible reaction mechanism for the formation of ZnO thin films is as follows [10]:

$$Zn\,(NO_3)_2 + 2NH_4OH \rightarrow Zn\,(OH)_2 + 2NH_4NO_3 \tag{1}$$

$$Zn\,(OH)_2 + NH_4OH \rightarrow Zn\,O_2^- + H_2O + H^+ \tag{2}$$

When the solution is heated to 353 K, the ionic product exceeds the solubility product, and precipitation occurred on the glass substrate and in the solution to form ZnO nuclei; thus, ZnO film is formed on the substrate by the following reaction [10]:

$$(NH_4)Zn\,O_2^- + H^+ \rightarrow ZnO + NH_4OH \tag{3}$$

The growth process of Cu-doped ZnO proceeded in a similar manner as that of undoped ZnO since doping is a physical process and does not significantly affect the chemical structure of the parent precursor, ZnO.

3.2 Structural characterization

Their CZO electrodes were characterized by X-ray diffraction which revealed that all the film samples were polycrystalline with hexagonal wurtzite structure with lattice constants a = 3.24982 Å and c = 5.20661 Å corresponding to those of the ZnO patterns from the JCPDS data card no: 00–036-1451. A strong preferential growth was seen along the Z-direction for the undoped ZnO and CZO films with 3% Cu concentration, which diminishes significantly for 1 and 5% Cu concentrations as illustrated in **Figure 6**. The decrease in intensity of the ZnO peaks upon incorporation of Cu impurities (i.e. Cu/Zn of 1%) was assigned to induce crystallographic defects on ZnO lattices by Cu dopant which reduced the crystalline quality of the film. They observed on the other hand that raising Cu content to 3% enhances the carrier concentration and mobility in the conduction band of the semiconductor [10], thereby lowering the amount of crystallographic defects in the film and increasing the crystalline quality of the film along the ZnO (002) plane [13]. Further increase in Cu content to 5% lowered the crystalline quality of the film sample significantly as indicated in **Figure 6**, which is an indication of more compressive strain in the films at higher doping level [1, 2].

Barna and Adamik structure zone model for polycrystalline metallic films [23, 24] explained the above-described phenomenon; thus as the Cu content in ZnO is raised, segregation occurred at the grain boundaries in the film, which give rise to shrinkage of crystallite sizes [23]. Therefore, 3% Cu concentration could be the optimum concentration of Cu in ZnO to provide the best crystalline quality film in CZO thin films. This observation slightly defers from an earlier report of Babikier et al. and Mkawi et al. [22, 25] even though they used different concentrations of Cu.

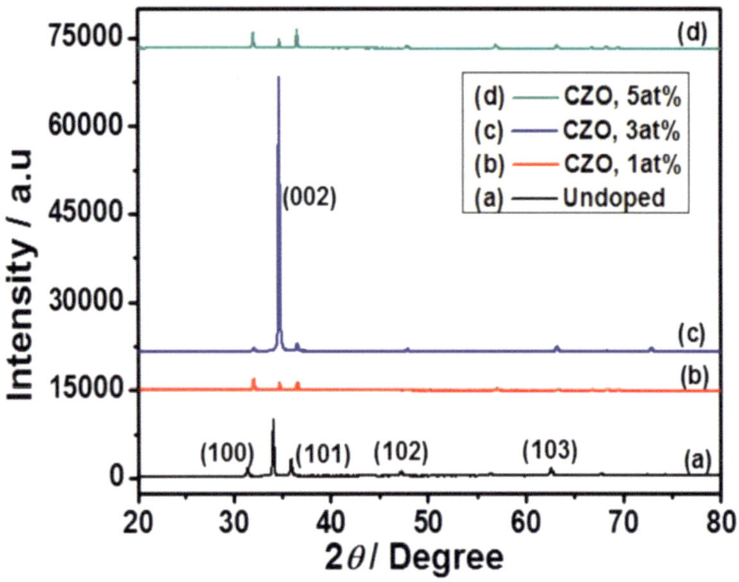

Figure 6.
XRD patterns of CZO with varying Cu concentrations: (a) 0%, (b) 1%, (c) 3% and (d) 5% [10].

Another interesting observation here is a shift in angular peak positions of 0.54° in (2θ) along the (002) plane in ZnO when Cu impurities were introduced. This may be assigned to difference in the ionic radii of Zn^{2+} and Cu^{2+} (which is higher for Zn^{2+} as compared to Cu^{2+}) as Cu^{2+} substitutes Zn^{2+} on Zn sites in ZnO, the crystal lattice, thus causing shortening of the c-axis [23]. According to Shannon [24], the fourfold coordinated Zn^{2+} and Cu^{2+} cations have ionic radii of 0.074 and 0.057 nm, respectively, and stable electronic configurations of Zn^{2+} (3d10) and Cu^{2+} (3d9). The fourfold coordinated Cu^{1+} has ionic radius of 0.06 nm Cu1(3d10) [24].

The mean crystallite sizes, *D* of the CZO thin films along the c-axis (002), were estimated on the basis of full width at half maxima (FWHM) using Scherrer's formula [25]:

$$D = \frac{0.9\lambda}{\beta Cos\theta} \tag{4}$$

where λ, β and θ are the X-ray wavelength (λ = 1.54 Å), full width at half maximum and diffraction peak angle, respectively. Their estimated crystallite sizes indicated a slight increase from 28 to 30 nm as depicted in **Table 1**.

In chemically synthesized ZnO thin film, as-deposited films may likely contain hydroxide and other impurities [25], so thermal annealing is necessary. Thermal annealing causes remarkable changes in the surface morphology of chemically synthesized thin films. Tyona et al. [10] reported the surface morphology of their

% Doping of Cu	Mean crystallite size (nm)
0	28
1	7
3	30
5	8

Table 1.
Estimated crystallite sizes of CZO thin films using (002) crystal plane [10].

CZO thin films which was studied using scanning electron microscope, JEOL JSM-6360. **Figure 7a–d** depicts the SEM micrographs, with undoped ZnO (as reference) and CZO thin films (as-deposited and annealed) with different Cu concentrations

Figure 7.
SEM micrographs of CZO thin films showing as-deposited and annealed samples for various Cu contents. (a) 0%, as-deposited and annealed. (b) 1%, as-deposited and annealed. (c) 3%, as-deposited and annealed. (d) 5%, as-deposited and annealed [10].

(0, 1, 3 and 5%). The SEM micrographs showed that the surface morphology of the films was affected by the concentration of the dopant. The crystallite size of the films decreases upon introduction of the Cu impurity as suggested by XRD study except for 3%. This behaviour could also be assigned to the difference in the ionic radii of Zn and the doping element, Cu. Further, with increasing Cu concentration (up to an optimum concentration of 3%), the microstructures of the film became denser as in **Figure 7c**. **Figure 7a(i)** and **(ii)** illustrates SEM micrographs of the undoped ZnO thin films (as-deposited and annealed) identified with randomly oriented nanorods of average diameter approximately 125 nm.

3.3 Surface morphological studies

The micrograph of the as-deposited CZO (i.e. 1%) film as depicted in **Figure 7b(i)** showed surface morphology with well-defined nanorods grown randomly on the substrate as thin solid films of Cu-$Zn(OH)_2$ with sparsely distributed needle-like images attached to some rods. Each rod is crystalline and indexed to hexagonal crystal structure, as seen in **Figure 7b(i)**. The average diameter of the rods was of the order of ≈40 nm. After annealing at 673 K, the nanorods.

Were refined to CZO with nanorod morphology of average rod diameter of ≈30 nm and randomly oriented, leading to large surface area as depicted in **Figure 7b(ii)**. Such novel morphology may find applications in photoelectrochemical solar cells, gas sensors and super capacitors [22].

Figure 7c(i) and **(ii)** represents the SEM micrographs of as-deposited and annealed CZO with Cu:Zn of 3%. The as-deposited film sample is identified with dense and vertically aligned nanorod morphology of varying rod sizes with average rod diameter of 120 nm and high porosity as well as high surface roughness. After post-annealing treatment at 673 K, the former nanorod morphology became fibrous, vertically aligned with well-defined and nearly uniform rod sizes (mean rod diameter of ≈112 nm) as depicted in **Figure 7c(ii)**. This morphology is suitable for dye-sensitized solar cell (DSSCs) application.

The SEM micrographs of CZO with Cu concentration of 5% are shown in **Figure 7d(i)** and **(ii)**. As-deposited film samples, d(i), show densely oriented nanorods. After post-annealing treatment at 673 K, dense, uniformly oriented nanorods with fine structures were obtained as in d(ii). These observations agree with that observed by Chow et al. [2] using chemical synthesis of CZO thin films and are suitable for PEC solar cell application.

3.4 Optical characterizations

The UV–Vis absorbance spectra measured in the work of Tyona et al. [10] is illustrated in **Figure 8**. The measurements were carried out in the wavelength range of 300–800 nm at room temperature, with undoped ZnO as reference. The spectra generally revealed low absorbance for all the film samples in the visible region especially the undoped ZnO and CZO samples with Cu concentrations of 1 and 5%, with harp absorption edge at about 395 nm. An enhancement in absorption was noticed in CZO with Cu:Zn of 3%; the band edge was shifted towards lower energy at 450 nm (**Figure 8**). This is an enhancement in optical absorbance induced by Cu doping.

The optical band gap energy (E_g) was obtained from Tauc plot using Tauc's relationship between the absorption coefficient, α, and the photon energy, $h\nu$, as shown in Eq. (5) [10]:

$$\alpha = \frac{\alpha(hv - E_g)^n}{hv} \qquad (5)$$

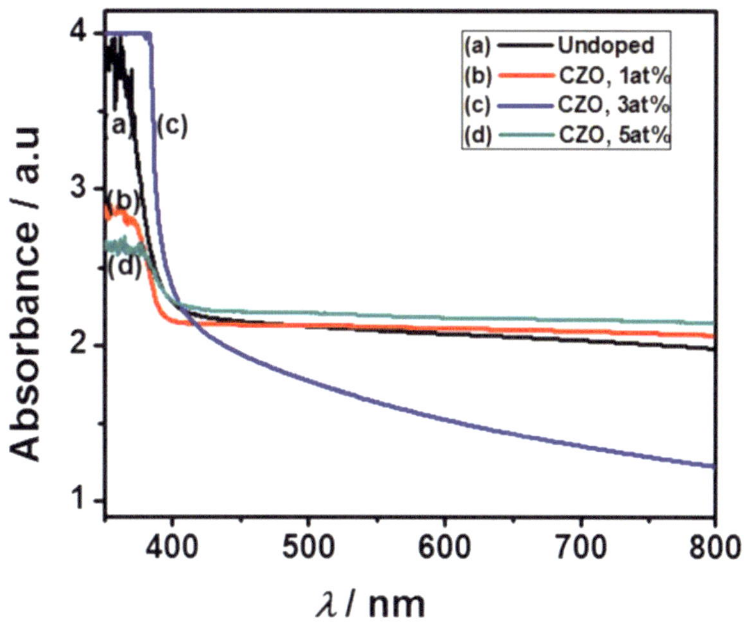

Figure 8.
Absorption spectra of CZO thin film for different Cu concentrations: (a) 0%, (b) 1%, (c) 3% and (d) 5% [10].

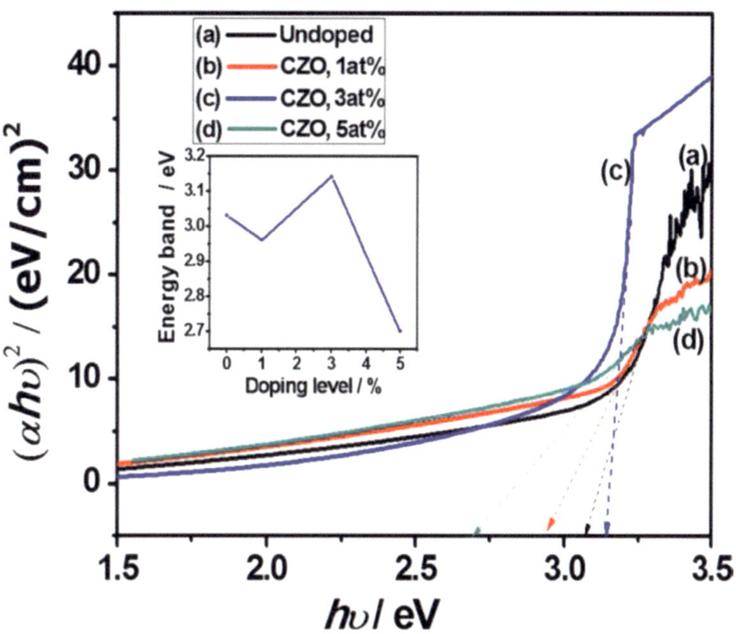

Figure 9.
Tauc plot from absorbance spectra of CZO thin film for different Cu concentrations: (a) 0%, (b) 1%, (c) 3% and (d) 5%. Inset shows the band gap trend of undoped and Cu-doped ZnO thin films [10].

where α_o is a constant. E_g is the optical band gap and n is a constant which depends on the probability of transition (it takes values as 1/2, 3/2, 2 and 3 for direct allowed, direct forbidden, indirect allowed and indirect forbidden transition, respectively) [22]. The band gaps of the CZO thin films were estimated by extrapolating to the photon energy axis, the linear portion of Tauc's plot as depicted in **Figure 9**.

The estimated E_g values of CZO thin films are observed to decrease from 3.03 eV for undoped ZnO to 2.96 eV upon incorporation of Cu impurities into ZnO (1%) and later increased to 3.14 eV as the impurity concentration was raised to 3% (in set, **Figure 9**). This anomaly in band gap of CZO thin films they observed was adduced to stress incorporated in the films which resulted from some defects on Zn sites induced by Cu impurities [22]. The reduction in the band gap observed thus facilitates excitation of electrons from the valence band to conduction band, even under illumination with visible light photons. This is beneficial for achieving an improved PEC performance from an UV-active material [19]. This study confirms that the incorporation of an optimum concentration of Cu in ZnO can yield a modification in the band gap of ZnO and thus provide an efficient solar light absorbing optical system, which can be useful for the solar energy applications [26].

Optical transmittance showed variation over a wavelength range of 350–800 nm. Regardless of the Cu concentration, all films are highly transparent in the visible (400–800 nm) region. The transmittance of Cu-doped films showed moderately high values in the range of 60–90% in the visible region as shown in **Figure 10**. Upon incorporation of Cu impurities, the optical transmittance was found to decrease from 85% (undoped) to 78% (CZO, 1%); further increase in Cu concentration enhanced the transmittance value to 90%. At higher doping concentration (above 3%), the transmittance was observed to decrease with increased Cu concentration which may be adduced to decrease in crystalline quality of the Cu-doped film.

3.5 Surface wettability studies

Surface wettability involves the interaction between liquids and solids in contact. The wetting behaviour of thin film is characterized by the value of contact angle, a microscopic parameter. The contact angle is an important parameter in surface science, and its measurement provides a simple and reliable technique for

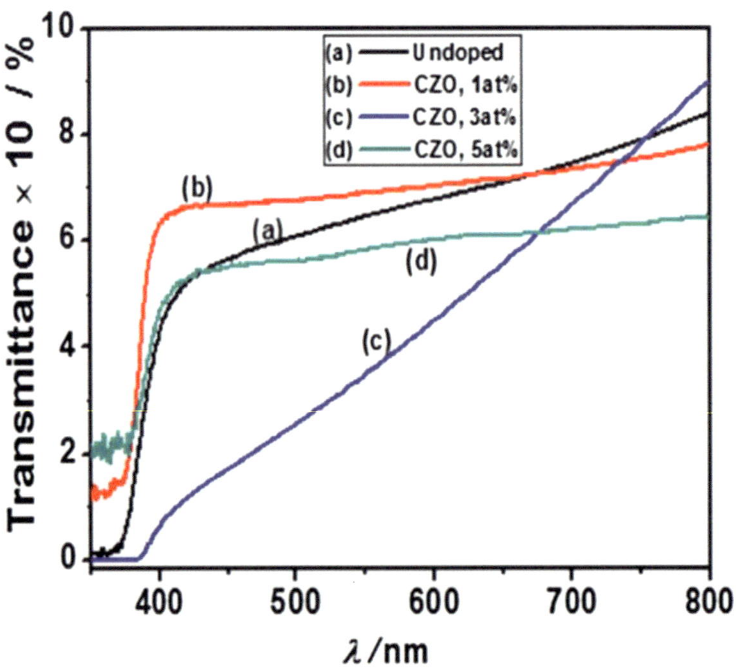

Figure 10.
Transmittance spectra of CZO thin films with different Cu concentrations: (a) 0%, (b) 1%, (c) 3% and (d) 5% [10].

the interpretation of surface energies [10, 27]. The knowledge of contact angle of thin films is important for practical applications such as in DSSCs, super capacitors and gas sensing [10, 28].

Tyona et al. [10] reported water contact angles of CZO thin films as illustrated in **Figure 11a–d**. The contact angles are observed to decrease from 71.3 to 15.2° upon doping. These results showed strong dependence of the contact angles on annealing temperature as well as Cu concentration. The water contact angles decreased (as compared with undoped sample **Figure 11b**) upon incorporation of Cu impurities up to optimal percentage impurity of 3% (**Figure 11c**) and further increase as the impurity concentration was increased beyond this level (**Figure 11d**). This implies

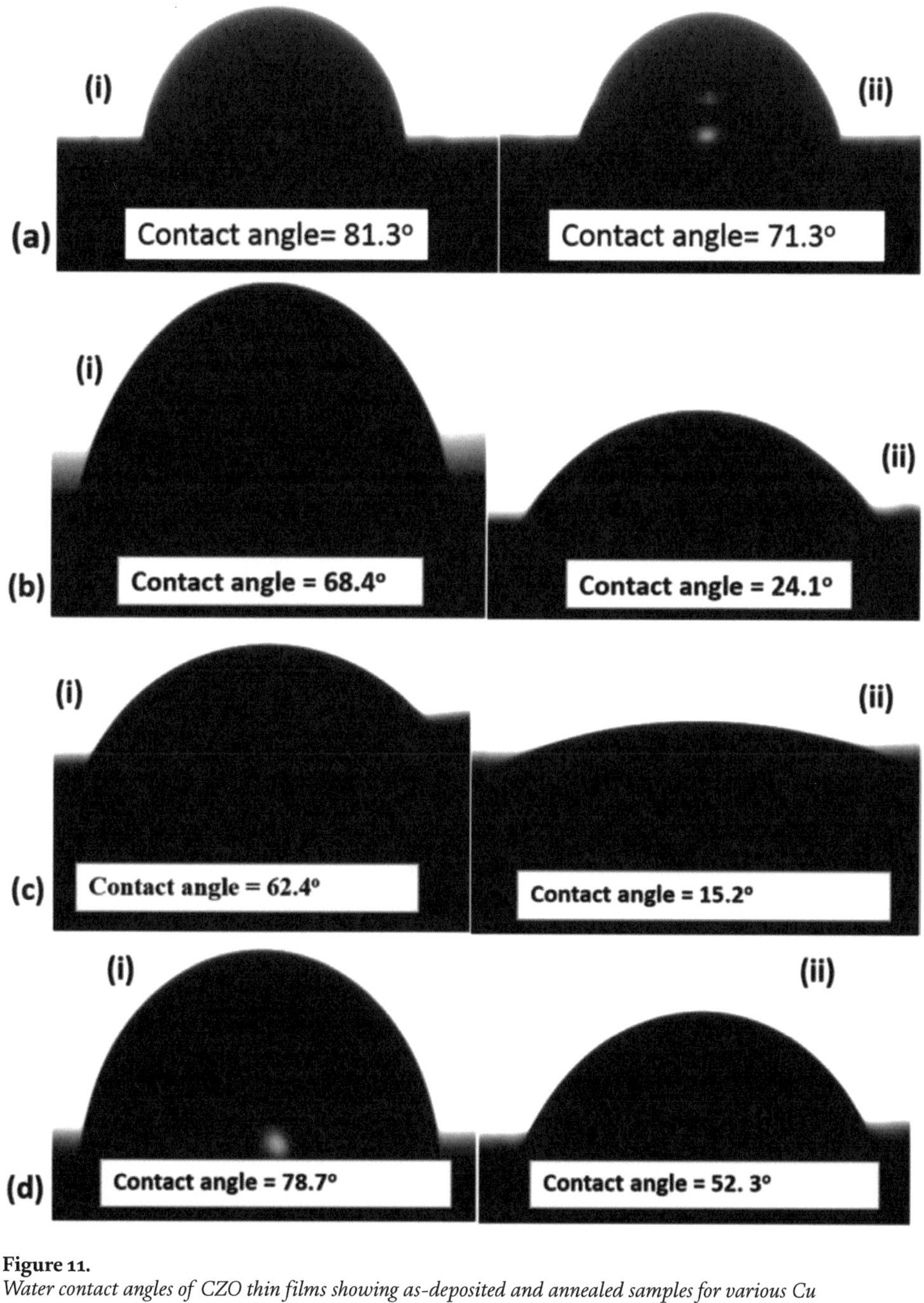

Figure 11.
Water contact angles of CZO thin films showing as-deposited and annealed samples for various Cu concentrations: (a) 0%, (b) 1%, (c) 3% and (d) 5% [10]. In **Figure 11***, (i) is as deposited, (ii) is annealed.*

that the porosity of the films increases upon thermal annealing at 673 K and with respect to increased concentration of Cu; thus, the water goes in to the pores and craves making contact angle hydrophilic [10, 28]. This means that the films would have large surface areas which for application such as DSSCs would mean better dye adsorption resulting to enhanced photo absorption. Also, lower values of contact angles are beneficial for electrolyte percolation through the porous film, which is very important for PEC solar cells. This result is in agreement with the earlier observations from XRD and SEM.

3.6 Photoelectrochemical (PEC) studies

Photoelectrochemical response of a solar cell is based on the junction between semiconductor and an electrolyte. The electrolyte plays an important role in PEC cell as a medium for charge transfer between the photoelectrode and counter electrode [10, 22].

The photoresponses of the CZO thin films were studied by forming typical configuration cells, n-CZO (stainless steel substrate)/0.1 M Na_2SO_4/platinum/SCE. These PEC cells are easy to form, and many processing steps of p–n junction have been simplified or eliminated. Since the junction with liquid is formed spontaneously upon contact, irregular-shaped single crystal or thin films can be used [10, 17]. The solution-based measurements allowed us to quickly test the quality of CZO film electrode as a solar cell material [10, 17].

Tyona et al. [10] reported the PEC performance of their CZO using current–voltage (I-V) characteristics of the annealed CZO thin films in the dark and under illumination with 80 mW/cm^2 as illustrated in **Figure 12a–c**. The anodic

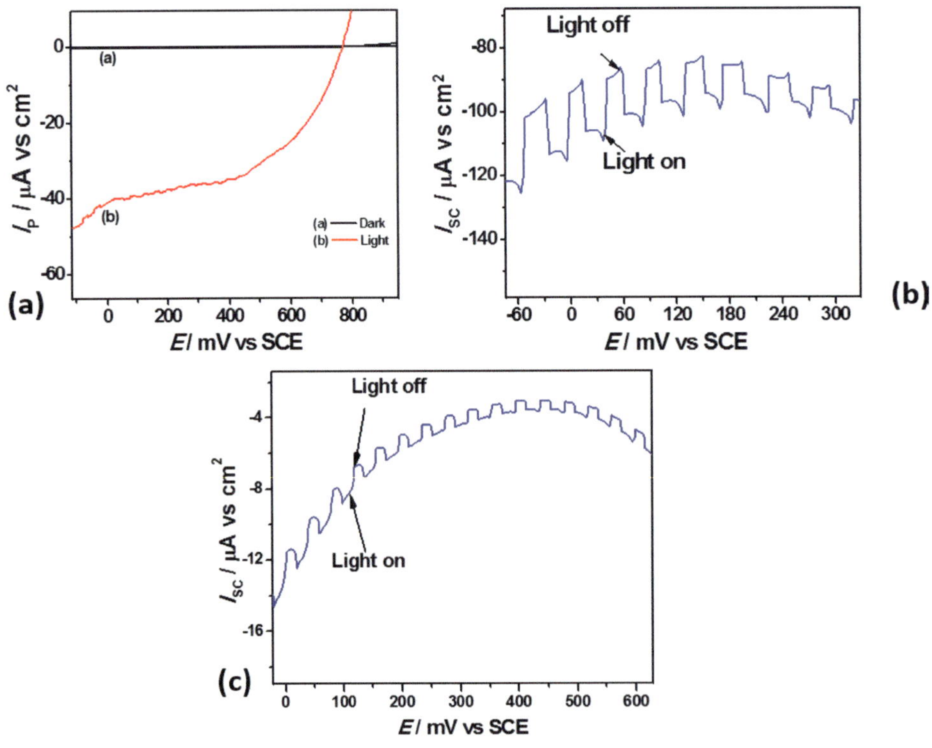

Figure 12.
Current-potential (I-V) curves of CZO thin films showing current and potential in the dark and under illumination for (a) 1%, (b) 3% and (c) 5% Cu concentrations [10].

ZnO electrodes	Photocurrent (I_{SC}) [μA/cm²]	Photovoltage (V_{oc}) [mV]	I_{max} (μA/cm²)	V_{max} (mV)	Efficiency η (%)	Fill factor (FF)
Undoped	12.34	388.0	9.00	230.0	0.0030	0.43
CZO (1%)	40.00	774.0	28.00	631.0	0.0220	0.57
CZO (3%)	98.00	796.0	74.00	667.0	0.0620	0.63
CZO (5%)	16.00	768.0	13.00	52.0	0.0009	0.06

Table 2.
Values of I-V measurement from PEC cells of CZO thin films [10].

photocurrent varied remarkably with the concentration of Cu in ZnO as reported with previous characterizations of the CZO. **Figure 12b** and **c** also represents the chopped light tests carried out in order to study the photosensitivity of CZO thin films. The photosensitivity confirmed that CZO absorber is an n-type material and is useful for the solar cell [10, 17]. The measured values of the PEC parameters with respect to Cu doping are shown in **Table 2**.

The photoelectrochemical measurement confirmed good photoactivities of the annealed CZO films prepared from simple CBD method. It is however observed that the photocurrent (short circuit current, I_{SC}) conversion efficiency and fill factor of the CZO film for 5% Cu are relatively low. This may be due to more compressive strain in the films at higher doping level as earlier explained in XRD which probably leads to a less dense nanostructure as illustrated in **Figure 7d** and consequently low photoactivity. The photocurrent obtained in the present study is not useful for most practical applications requiring high values of current; however, it is well known that conversion efficiency of such film can be considerably improved by thermal, chemical and photoelectrochemical surface treatments [10, 17].

4. Conclusions

This chapter examines ZnO and its numerous nanostructures and also considered doping as a measure for engineering the properties of ZnO for pre-determined applications. The chapter has also extensively reviewed the effect of Cu doping on structural, morphological and optical properties and surface wettability of chemical bath deposited ZnO thin films at various concentrations of Cu in the range 1–5% for PEC solar cell application. The review indicated that there were slight changes in the lattice parameters of the CZO electrodes which occurred due to the successful substitution of Zn^{2+} by Cu^{2+} and also enhancement in crystalline quality of the films at 3% Cu concentration due to the reduction in crystallographic defects in the film. A review of SEM studies showed densely grown nanorods over the varied range of Cu concentration, with the CZO nanorods of 3% having the most dense microstructures with average diameter approximately 125 nm. The density and diameter of the nanostructures demonstrated dependence on the amount of Cu dopant. A review of optical properties demonstrated that the incorporation of Cu dopant into ZnO introduced a shift in absorption edge of approximately 60 nm into the visible band for the CZO nanorods with 3% Cu content which is a significant enhancement in the optical properties of the films. Also, optical energy band gaps decrease from 3.03 to 2.70 eV upon Cu doping. Surface wettability was adjudged hydrophilic for all the films, which implied high porosity, and the size of water contact angles show dependence on Cu content. Photoelectrochemical cell performance indicated an n-type photoactivity in sodium sulphate (Na_2SO_4) electrolyte which motivate to check its feasibility in solar cell applications.

Acknowledgements

I am grateful to Benue State University, Makurdi, for providing an enabling environment for this work.

Conflict of interest

I declare that there is no conflict of interest.

Author details

Tyona MD
Department of Physics, Benue State University, Makurdi, Benue State, Nigeria

*Address all correspondence to: dtyona@gmail.com; dtyona@bsum.edu.ng

References

[1] Zhong LW. Zinc oxide nanostructures: Growth, properties and applications. Journal of Physics. Condensed Matter. 2004;**16**:R829-R858

[2] Chow L, Lupan O, Chai G, Khallaf H, Ono L, Roldan K, et al. Synthesis and characterization of Cu-doped ZnO one-dimensional structures for miniaturized sensor applications with faster response. Sensors and Actuators A. 2013;**189**:399-408

[3] Drici A, Djeteli G, Tchangbedgi G, Deruiche H, Jondo K, Napo K, et al. Structured ZnO thin films grown by chemical bath deposition for photovoltaic applications. Physica Status Solidi (a) Banner. 2004;**201**:1528-1535

[4] Li Y, Gong J, Deng Y. Hierarchical structured ZnO nanorods on ZnO nanofibers and their photoresponse to UV and visible lights. Sensors and Actuators A: Physical. 2010;**158**:176-187

[5] Lao CS, Liu J, Gao P, Zhang L, Davidovic D, Tummala R, et al. ZnO nanobelt/nanowire Schottky diodes formed by dielectrophoresis alignment across Au electrodes. Nano Letters. 2006;**6**:263-275

[6] Tyona MD, Osuji RU, Ezema FI, Jambure SB, Lokhande CD. Enhanced photoelectrochemical solar cells based on natural dye-sensitized Al-doped zinc oxide electrodes. Advances in Applied Science Research. 2016;**7**:18-31

[7] Vanaja A, Ramaraju GV, Srinivasa RK. Structural and optical investigation of Al doped ZnO nanoparticles synthesized by sol-gel process. Indian Journal of Science and Technology. 2016;**9**:23-28

[8] Cebulla R, Wndt R, Ellmer K. Al-doped zinc oxide films deposited by simultaneous RF and DC excitation of a magnetron plasma: Relationships between plasma parameters and structural and electrical film properties. Journal of Applied Physics. 1998;**83**:1087-1095

[9] Seshan K. Handbook of Thin-Film Deposition Processes and Techniques: Principles, Methods, Equipment and Applications. Second ed. New York, U.S.A: William Andrew Publishing Norwich; 2002. pp. 344-356

[10] Tyona MD, Osuji RU, Asogwa PU, Jambure SB, Ezema FI. Structural modification and band gap tailoring of zinc oxide thin films using copper impurities. Journal of Solid State Electrochemistry. 2017;**21**:2629-2637

[11] Snure M, Tiwari A. Band-gap engineering of $Zn_{1<x}Ga_xO$ nanopowders: Synthesis, structural and optical characterizations. Journal of Applied Physics. 2008;**104**:073707-073705

[12] Singhal S, Kaur J, Namgyal T, Sharma R. Cu-doped ZnO nanoparticles: Synthesis, structural and electrical properties. Physica B. 2012;**407**:1223-1226

[13] Dom R, Lijin RB, Kim HG, Borse PH. Enhanced solar photoelectrochemical conversion efficiency of ZnO:Cu electrodes for water-splitting application. International Journal of Photoenergy. 2013;**2013**:9-20

[14] Zhou Z, Kato K, Komaki T, Yoshino M, Yukawa H, Morinaga M, et al. Electrical conductivity of Cu-doped ZnO and its change with hydrogen implantation. Journal of Electroceramics. 2003;**11**:73-79

[15] Jongnavakit P, Amornpitoksuk P, Suwanboon S, Ndiege N. Preparation and photocatalytic activity of Cu-doped ZnO thin films prepared by the sol-gel method. Applied Surface Science. 2012;**258**:8192-8198

[16] Tyona MD, Osuji RU, Ezema FI. A review of zinc oxide photoanode films for dye-sensitized solar cells based on zinc oxide nanostructures. Advanced Nano Research. 2013;**1**:43-58

[17] Shinde NM, Dubal DP, Dhawale DS, Lokhande CD, Kim JH, Moon JH. Room temperature novel chemical synthesis of Cu_2ZnSnS_4 (CZTS) absorbing layer for photovoltaic application. Materials Research Bulletin. 2012;**47**:302-307

[18] Machado G, Guerra DN, Leinen D, Ramos-Barrado JR, Marotti RE, Dalchiele EA. Indium doped zinc oxide thin films obtained by electrodeposition. Thin Solid Films. 2005;**490**:124-131

[19] Tyona MD, Jambure SB, Lokhande CD, Banpurkar AG, Osuji RU, Ezema FI. Dye-sensitized solar cells based on Al-doped ZnO photoelectrodes sensitized with rhodamine. Materials Letters. 2018;**220**:281-284

[20] Becerril M, Silva-López H, Guillén-Cervantes A, Zelaya-Ángel O. Aluminum-doped ZnO polycrystalline films prepared by co-sputtering of a ZnO-Al target. Revista Mexicana de Física. 2014;**60**:27-31

[21] Muthukumaran S, Gopalakrishnan R. Structural, FTIR and photoluminescence studies of Cu doped ZnO nanopowders by coprecipitation method. Optical Materials. 2012;**34**:1946-1953

[22] Babikier M, Wang D, Wang J, Li Q, Sun J, Yan Y, et al. Cu-doped ZnO nanorod arrays: The effects of copper precursor and concentration. Nanoscale Research Letters. 2014;**9**:199-207

[23] Thakur S, Sharma N, Varkia A, Kumar J. Structural and optical properties of copper doped ZnO nanoparticles and thin films. Advances in Applied Science Research. 2014;**5**:18-24

[24] Shannon RD. Revised effective ionic radii and systematic studies of interatomic distances in halides and chalcogenides. Acta Crystallogr. Sect. A. 1976;**32**:751-767

[25] Mkawi EM, Ibrahim K, Ali MKM, Farrukh MA, Mohamed AS. The effect of dopant concentration on properties of transparent conducting Al-doped ZnO thin films for efficient Cu_2ZnSnS_4 thin-film solar cells prepared by electrodeposition method. Applied Nanoscience. 2015;**3**:56-67

[26] Mani GK, Rayappan JBB. Influence of copper doping on structural, optical and sensing properties of spray deposited zinc oxide thin films. Journal of Alloys and Compounds. 2014;**582**:414-419

[27] Sun RD, Nakajima A, Fujushima A, Watanabe T, Hashimoto K. Photoinduced surface wettability conversion of ZnO and TiO_2 thin films. The Journal of Physical Chemistry. B. 2001;**105**:1984-1991

[28] Sun H, Luo M, Weng W, Cheng K, Du P, Shen G, et al. Room-temperature preparation of ZnO M nanosheets grown on Si substrates by a seed-layer assisted solution route. Nanotechnology. 2008;**19**:125603-125610

Section 4

ZnO Applications

Chapter 5

Pyrolysis of Carbon-Doped ZnO Nanoparticles for Solar Cell Application

Luyolo Ntozakhe and Raymond Tichaona Taziwa

Abstract

It is very important to find new methods for improving the properties of nano-structured materials that can be used to replace the highly expensive and complicated techniques of fabricating ZnO nano-powders for solar cell applications. Pneumatic spray pyrolysis method offers a relatively inexpensive way of fabricating ZnO nanomaterials of controllable morphology, good crystallinity and uniform size distribution, which makes it a good candidate for the production of ZnO nanoparticles. Additionally, it has the advantage of producing ZnO NPs in one step directly on the substrate without the need for other wet chemistry processes like purification, drying and calcination. To that end, the present study emphasizes more on the design and optimization of spray pyrolysis system as well as on the pneumatic spray pyrolysis conditions for the production of carbon-doped ZnO nanoparticles. The un-doped and carbon-doped ZnO NPs were prepared using pneumatic spray pyrolysis employing zinc acetate as a precursor solution and tetrabutylammonium as a dopant. The fabricated un-doped and C-ZnO NPs were characterized for their morphological, structural and optical properties using SEMEDX, XRD and DRS. SEM analysis has revealed that the fabricated un-doped and C-ZnO NPs have spherical shape with mesoporous morphology. The cross-sectional SEM has also revealed that the film thickness changes with increasing dopant concentration from 0.31 to 0.41 μm at higher concentrations. Moreover, the EDX spectra have confirmed the presence of Zn and O atoms in the PSP-synthesized ZnO NPs. XRD analysis of both un-doped and C-ZnO has revealed the peaks belonging to hexagonal Wurtzite structure of ZnO. Additionally, the DRS has revealed a decrease in energy band gap of the synthesized ZnO NPs, with the increase in carbon dopant level.

Keywords: spray pyrolysis, zinc oxide, nanoparticle, pneumatic spray pyrolysis

1. Introduction

In recent decades, semiconducting metal oxide materials such as zinc oxide (ZnO), tin oxide (SnO_2), iron oxide (Fe_2O_3) and titanium dioxide (TiO_2) have become an area of research due to their great potential to solve environmental problems [1]. These wide-band-gap semiconductors are considered to have the ability to easily adjust the optoelectronic and transport properties of the metal oxide semiconductor material which makes them the promising candidates for several applications such as gas sensing, photocatalysis, storage and solar energy conversion [2, 3]. Among them, zinc oxide is of special interest due to its wide range of properties such as direct band gap (3.37 eV at room temperature), non-toxicity, high photo stability and large exciton binding energy (60 meV) [3, 4]. Additionally, ZnO offers a low-cost material for solar cell electrodes, and its low price compared with other wide-band semiconductor nanomaterials makes it a good candidate for industrial applications.

Moreover, the physical and chemical properties of ZnO nanomaterial depend closely on the two geometrical parameters which are size and shape [5]. It can be easily processed into several nanostructures of different sizes, shape and morphologies such as nanorods, nanowires, nanonails, nanotubes, nanocombs, nanoflower, nanosheets, nanobelts, nanoparticles, etc. These nanostructures have become an area of interest for several applications due to their unique properties observed at nanoscale. Among the numerous nanostructures, nanoparticles (NPs) have attracted great attention in many device applications due to their unique catalytic, optical, magnetic and electrical properties because of their nanoscale dimensions [5, 6]. Particularly, the modified ZnO NPs have become function and integration of nanostructure assembly for dye-sensitized solar cells and nanoscale devices. The structural, morphological and optical properties of ZnO NPs can be improved and controlled by a chemical reaction which is a very crucial factor influencing the performance of semiconductor nanomaterials for many technological applications. Recently, great efforts have been made in the fabrication of doped ZnO nanomaterials in order to improve the morphological, structural and optical properties of nanomaterials specifically by modifying the surface properties such as electronic band gap, specific surface area, oxygen vacancies and crystal deficiencies [6, 7]. In previous articles, it has been reported that doping with non-metal elements such as C, S and N can improve the structural, morphological and vibrational properties of ZnO [7–9]. Doping with non-metal elements such as N or C has been reported to reduce the band gap of a wide-band semiconductor by enhancing a number of properties but not limited to ferromagnetism, magnets to transport properties and p-type conduction properties [8, 9].

There are several methods that have been used for the preparation of ZnO NPs which include physical methods and wet chemistry to solid-phase systems. However, the physical and chemical methods such as evaporation plasma [10], anodization [11, 12], spin on methods [13, 14], sputtering [15], ion-assisted deposition [16], reactive ion plating [17], laser ablation [18], filtered arc deposition [19] and atomic layer epitaxy [20] employed for the preparation of ZnO NPs in previous articles have been reported to require very expensive equipments, complex process controls and stringent reaction conditions. Additionally, these methods are responsible for the host of many problems such as generation of hazardous by-products, the use of toxic and flammable solvents, difficult controlling of morphology, very substrate dependent and require high vacuum temperature. The solid-based methods are the promising alternatives for the preparation of ZnO NPs due to their simplicity and high yield [9, 21]. Additionally, they do not require several steps like wet chemistry methods such as purification, calcinations, drying and extraction of material before use.

Moreover the solid-based techniques present no reagent concentration drawback as compared to other physical and chemical systems, which makes them more desirable for industrial scaling. Among the solid-based methods like laser ablation [21], sputtering coating [22], frequency electron tube sputtering (RFMS) [23], spin coating [24], aerosol-assisted chemical vapor deposition [25] and spray pyrolysis [26]. Spray pyrolysis (SP) offers a modest and cost-effective way of fabricating a number of nanostructures unlike many other film deposition techniques [27].

Moreover, SP has the potential to produce nanoparticles with consistent structural, optical and morphological properties in a one-step process without a need for other wet chemistry steps such as cleansing or excessive drying that contributes on the overall cost of fabrication of ZnO nanostructures [27, 28]. Additionally, this method is responsible to coat huge substrates and also offers the possibility for industrial scaling. Furthermore, SP process makes it possible to grow nanoparticles directly on the substrate and can be used directly as photo-anodes in the dye-sensitized solar cell device which in turn lowers the overall production cost of a solar cell device. The present study focuses on the novel pyrolysis of carbon-doped ZnO nanoparticles for solar cell application. Additionally, the death of scientific articles that report on the influence of carbon doping on ZnO NPs fabricated by PSP method was also the major motivation of the present compilation.

2. Design consideration of a pneumatic spray pyrolysis system

2.1 Experimental section

ZnO nanoparticles were prepared by spray pyrolysis technique employing zinc acetate (Zn $(O_2CH_3)_2$, Merck), ethanol (99.99% Sigma Aldrich), acetic acid (CH_3COOH) and tetrabutylammonium bromide (TBA) $(C_{16}H_{36}BrN)$ as the starting materials. In typical experimental procedure, 0.1 M of zinc acetate was prepared by dissolving 5.4923 g Zn $(O_2CH_3)_2$ to a 250-ml volumetric flask containing minimum amount of ethanol, which results to the formation of zinc ethoxide solution. Then few drops of acetic acid were added to the zinc ethoxide (Zn $(O_2CH_3)_2$) solution as a stabilizer. The zinc ethoxide solution was sonicated inside the ultrasonicater for 30 min at 40°C. Tetrabutylammonium bromide about 1.55 ml was added to the resulting colorless solution of Zn $(O_2CH_3)_2$ in the volumetric flask. The 250-ml volumetric flask containing Zn $(O_2CH_3)_2$ and $(C_{16}H_{36}BrN)$ was filled up to the mark with absolute ethanol. Several other carbon dopant $(C_{16}H_{36}BrN)$ precursor solutions were prepared in much the same way as this one; the only difference was the volume of the dopant solution as shown in **Table 1**.

The precursor solution was then transferred into a chamber connected to a pneumatic pump prior to spray deposition. Firstly, the $F{:}SnO_2$ glass substrate was washed with detergent and then rinsed with distilled water, isopropanol, distilled water and acetone, followed by drying under hot air to evaporate the acetone. Then the fabrication of un-doped and C-ZnO NPs was done on well-dried $F{:}SnO_2$ glass substrate at a deposition temperature of 400°C. The fabricated NPs were then characterized using XRD spectra, obtained using a Bruker D8 Advance X-Ray diffractometer (XRD) with a Cu anode, generating Kα radiation of wavelength 1.544 Å and operating at 40 kV and 40 mA, which was used to obtain the crystallographic phase and associated parameters of ZnO samples of the fabricated NPs. The XRD θ-2θ patterns of ZnO NPs fabricated by PSP were recorded in the 2θ range of

Sample #	Mass of zinc acetate (g)	Volume of TBA (ml)	Millimoles of TBA
Un-doped	5.4923	0	0
0.010 M	5.4925	1.55	2.51
0.015 M	5.4922	2.33	3.78
0.025 M	5.4924	3.88	6.30

Table 1.
Preparation of precursor of solutions.

30–100°C at room temperature. The elemental, structural and morphological properties of ZnO nanostructures were studied using field emission scanning electron microscope (FE-SEM) Zeiss Auriga SEM equipped with EDS with Smart SEM software at an accelerating voltage of 30 kV. High-resolution transmission electron microscope was used to determine the finer details relating to structural properties of un-doped and C-ZnO samples, obtained using a JEOL JEM 2100 HRTEM operating at 200 kV for high-resolution images with selected area electron diffraction (SAED) patterns. The confocal Raman imaging system (WITec GmbH, Ulm, Germany) alpha300RS was used to study the Raman active modes present in the ZnO samples. A fibre-coupled DPSS laser 532 nm with maximum output power after single-mode fibre coupling of 44 mW was used as the excitation source. Data were then collected using a multimode fibre into a high-throughput lens-based spectrograph (UHTS 300) with 300 mm focal length and two gratings 600 and 1800 g/mm, both blazed at 550 nm.

2.2 Experimental problems experienced with spray pyrolysis system at other laboratories for production of ZnO nanomaterials

This part outlines the practical knowledge or challenges that other authors experienced when operating spray pyrolysis (SP) technique. The knowledge collected from these authors was considered and played an imperative part in governing, assembling and monitoring the current spray pyrolysis (SP) system; hence we are also working with similar spray pyrolysis system. Additionally, the knowledge assembled assisted to shape several considerations in selecting the equipment such as furnace type, pneumatic pump size, reactor type, as well as type pneumatic vessel of the existing SP system. Lastly, the deposition conditions as well as material properties employed to enhance the design and development of the pneumatic spray pyrolysis system were also considered.

3. Experimental complications

3.1 Use of zinc acetate precursor solution

In this study, spray pyrolysis deposition was operated on the glass substrate lying perpendicular on the floor of aluminum tube inside the furnace. The zinc acetate was used as a precursor solution in this research. The spray pyrolysis system used in this research employed pneumatic pump which helps to blow the precursor solution to form mist of droplets inside the PSP vessel. There were few problems occurred with regards to formation of ZnO NPs such as the low solubility of zinc acetate salt in ethanol which needed to be heated up in order to dissolve completely in the solvent and the aluminum contaminations (impurities) initiated from the aluminum tube that was used as the reaction vessel. At higher temperatures, approximately 400°C and above, the aluminum tube impurities penetrate from the sample in the substrate holder which caused sample contamination. Hence, the fabricated ZnO nanoparticles were contaminated with aluminum. Due to aluminum impurities found in the fabricated ZnO NPs, there is a necessity to develop a tube reactor that can withstand deposition temperature of approximately 400°C and above, which does not introduce any impurity element in the fabricated nanomaterials. Similar problems were found in the literature in the case of TiO_2 nanoparticles where titanium tetrachloride was used as a precursor solution which generated massive complications with its chloride contaminants. It was observed that the cubic structures of sodium chloride crystallize fast and obstruct the

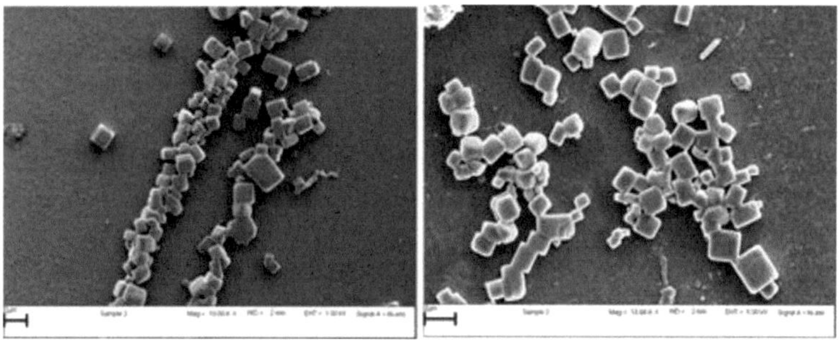

Figure 1.
SEM micrograph presenting the formation of NaCl crystals on top of the FTO glass substrate.

development of TiO_2 nanostructures on the uppermost of the glass substrates. Therefore, the contaminants of sodium initiated from the reactor tube that was used in the spray pyrolysis system. At higher temperature above 400°C, the sodium contaminant ions penetrate from the reactor tube which created nanomaterial impurities and reserved the development of TiO_2 nanomaterials on top of the glass substrates. **Figure 1** reveals the scanning electron microscopy image of cubic sodium chloride crystals on top of fluorine-doped tin oxide (FTO) glass substrates.

The zinc acetate precursor solution fumes are very toxic and acidic and have rapid diffusion rates at higher temperatures. The generated vapors were extremely acidic and portable at higher temperatures. Zinc acetate is known to be non-toxic, so these problems may be caused by the presence of acetic acid stabilizer in zinc acetate solution. The developed fumes produced almost rusted all the rubber tubing and seals fitted on the PSP system which may probably cause serious leakage problems in the system. Moreover the zinc acetate precursor vapors that are generated through deposition method have reserved the formation of zinc oxide nanoparticles; that is why it took us more hours to deposit a thin film of desirable thickness of 10 μm. This can also be probably due to low atomizer comprising the zinc acetate precursor capacities in the PSP vessel.

3.2 Substrate orientation

It was discovered that the thickness of the deposited ZnO thin film were not the same in all areas in the substrate, in some areas ZnO thin film had thicker layers compared to others and in some regions there were no ZnO NPs found completely as evidenced by **Figure 2**. It is noticed that only those regions of the substrates opposite the incoming vapor (aerosol) were completely covered. Then in other parts of the substrates such as those down the incoming vapor were moderately

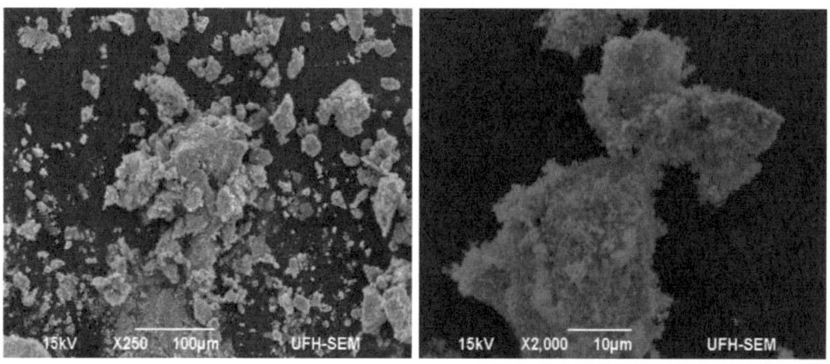

Figure 2.
SEM micrograph of non-uniformly coated ZnO thin film.

treated due to the position of the substrate (substrate orientation). The fabricated nanoparticles were found scattered in some areas of the substrates; this is probably due to poor substrate orientation or to unexpected temperature decrease when the furnace was turned off.

Moreover, the system took several hours to calm down to room temperature which made it difficult to run two or more samples a day, and the furnace used had poor heat insulation which dissipated heat into the working environment which made it very difficult to deposit thin films continuously. It was also discovered that the spray pyrolysis system has to cool down for more than 8 hours; otherwise, when quickly running a sample before 8 hours of cooling, the system just breaks down and it is very expensive to repair. Additionally, it was also noticed that after several hours of operation, the temperature rises up to beyond human comfort levels, which provided some difficulties in monitoring the spray pyrolysis deposition process continuously as the room temperature rises up to beyond 37°C. Hence it was very difficult to operate this kind of SP system right on a standard laboratory worktable because it presents a very high fire risk. Furthermore, all the rubber tubings could not resist the high temperatures and are needed to be changed regularly from time to time. This presented other possible health risks due to leakage of the evaporating zinc acetate precursor solution into the working atmosphere.

4. Pneumatic spray pyrolysis operation conditions

4.1 Safety

The pneumatic spray pyrolysis system needs to be closed at all times when running samples so it can strongly function on a normal laboratory workbench without any damage. The PSP system also needs to have an exhaust pipe into a suitably examined area to avoid exhaust of gas products into the work environment which are known to cause health problems in the long term. It has been reported that excessive inhalation of zinc acetate fumes causes nausea, diarrhea, metallic taste, kidney problem, stomach damage and vomiting. The tube furnace does not monitor the heat very well compared to split tube furnace so it was necessary to change the tube furnace by split tube furnace which can easily control the heat.

This has made it very simple and safe to work with aluminum tubing as the reaction reactor compared to quartz tubing. The split tube furnace was constructed in a manner that there is no heat that dissolute inside the work atmosphere. There was a necessity to use zinc acetate as a precursor solution since it is a safe, non-toxic liquid and can also prevent the requirement for high-priced gas handling method. Other precursor solutions like zinc chloride are not safe to use in the spray pyrolysis system because of the inherent problems of faster crystallization rate of chloride impurities which inhibits the formation of ZnO nanoparticles on the substrates. However, other salts like zinc nitrate and zinc naphthenate were other attractive electives for precursor materials for preparation of ZnO nanoparticle due to their safety, non-toxicity and their potential to produce well-adherent and uniform ZnO film.

4.2 Pneumatic spraying vessel

The main aim of this study was to design a pneumatic spray pyrolysis system that is capable of generating nanomaterials on a glass substrate, which can be compared to other previous findings by Mwakikunga et al. [29] and Taziwa et al. [30] using an ultrasonic reaction method consisting of an exciting capacity of 0.1 m^3.

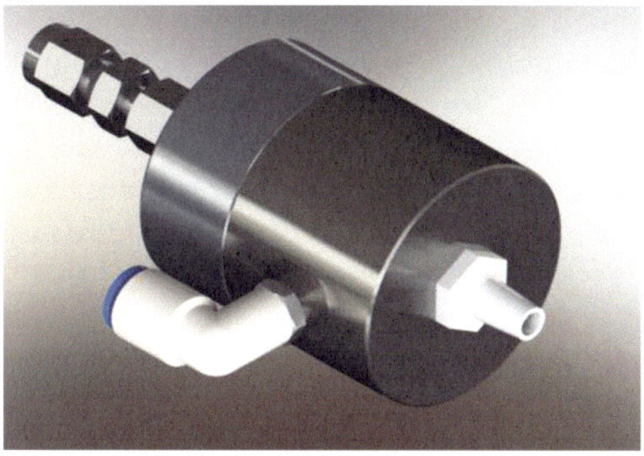

Figure 3.
Image of the PSP reaction vessel used in this study [31].

Figure 3 displays a PSP reaction vessel employed for fabrication of ZnO nanomaterials in this study. The precursor solution of zinc ethoxide is blown by the air from the pump to form a mist of droplets inside the PSP vessel. The material employed for the container presented a non-corrosive atmosphere for the accommodation of ZnO precursor solution.

4.3 Selection of ZnO precursor

The selection of a precursor solution became one of the most imperative methods in the fabrication of ZnO nanostructures using spray pyrolysis technique. The properties (physical and chemical) of the precursor solution are influenced by the solvent chosen, nature of salt, the salt concentration as well as other additive or extracts involved [32]. Hence, the structural, morphological and optical properties of the deposited thin films are easily formed through altering the nature of the precursor solution. The effects of preparation technique and zinc oxide precursor solutions like zinc nitrate and zinc chloride were studied. It has been reported that zinc chloride results in the formation of several unknown impurities in the sample which requires it to undergo various steps of wet chemistry such as washing, drying and calcination. Additionally, the samples originating from zinc chloride in the presence of sodium hydroxide as the precursor solution always produce the hexagonal Wurtzite structure of ZnO, while samples produced from the aqueous solution of zinc nitrate mostly result in the formation of polycrystalline structure of ZnO.

The preparation of ZnO using other methods of synthesis such as hydrolysis of zinc acetylacetonate monohydrate or hydrolysis of zinc naphthenate at room temperature also results in the formation of the hexagonal Wurtzite structure of ZnO. In the present study, the zinc acetate was chosen as the precursor of choice as the ZnO precursor. Despite the fact that it is one of the mostly used precursors in previous studies, it is a good candidate for solar cell production because the zinc acetate precursor can be easily modified with other additives like acetic acid as well as dopant to improve the resulting ZnO properties. Many published articles in the literature have highlighted that the morphologies of the fabricated nanostructures can be successfully modified by introducing preservatives into the precursor solutions [30]. Introduction of additives like CH_3COOH (acetic acid) in the precursor solution results in the alteration of the structure of the deposited ZnO films from fractured to a fractured free reticular. The change in the morphology of the deposited nanoparticles can be accredited to the change in precursor solution interaction. Hence in this work, zinc acetate was employed as the precursor for the fabrication of ZnO nanostructures. Moreover, the precursor consists of following advantages:

(1) non-corrosive and non-hazardous, recorded as a slight film, (2) extremely filtered and has nearly unlimited shelf life, (3) very unstable at slight temperatures such as 50°C, which indicates that it can be freely decomposed, (4) simply sprayed true without reduction, (5) it is relatively easy to handle as a liquid, even if it can be visible to an unmasked flame. The indication that it is not hazardous implies that presence of zinc acetate system is a relatively easy and safe task, as no special gas handling equipment is required, (6) it has been observed that it cannot be easily affected by the presence of oxygen in the atmosphere; hence there was no carrier gas used during the synthesis process.

4.4 Geometry mechanism

Another important aspect that required special consideration was the matter of substrate location and orientation in the reactor tube inside the furnace. It was almost impossible to deposit ZnO NPs at a position of 30° angle or with a substrate placed on the surface of the aluminum tube as this has been reported to cause inhomogeneous thin film coating. To avoid inhomogeneous thin film coating, there was a need to select or design an aluminum substrate holder that preserves the substrate at an angle of 90° so that it can directly interact with the incoming aerosol. The aluminum was selected as a metal of high quality since it presents several advantages such as its stability at deposition temperatures between 400 and 500°C, excellent conductor of heat and it allows preheating of the substrates before the deposition of thin film.

4.5 Deposition area

Almost all the types of solar cells have an active area where the semiconductor nanomaterials are deposited in the glass substrate and the rest of the area on the glass substrate can be utilized for solar cell contacts. Hence, it is necessary to mask the glass substrate in order to allow the formation of a thin film size of 0.5 cm^2 that is normally required for dye-sensitized solar cell application. Hence a substrate holder was designed in such a way that it could stand inside the aluminum tube reactor and at the same time other contact areas were masked during deposition.

4.6 Selection of spray reactor (split tube furnace)

There are several methods that can be used to synthesize the ZnO nanomaterials including wet chemistry techniques and solid-based methods. The solid-based methods are the most preferred for producing nanomaterials compared to solution-based methods due to their simplicity, high yield and they do not require several steps of wet chemistry such as washing, drying as well as calcination. The spray pyrolysis system that employ tubular reactor systems (furnace) offer several advantages such as (a) operation simplicity, (b) minimum throughput, (c) smaller current budget, (d) inexpensive to operate and (e) does not require successive heat treatment just after sample synthesis. Additionally, when tubular reactor is employed in spray pyrolysis method, each droplet holds the precursor in the exact the same stoichiometry as preferred in the product unlike in other reactors such as vapor flame reactor where the variety of the products is limited by the choice of metal precursors with enough vapor pressure to provide the preferred amount of the species into the reactor. Moreover, in spray pyrolysis reactor, the droplet basically serves as an isolated micro-reactor which is a vast advantage over the vapor phase method because

Figure 4.
Image of the split tube furnace and aluminum reactor used in the fabrication of ZnO nanostructures.

most reactants are expected to evaporate concurrently with extra precaution to achieve the preferred stoichiometry in vapor flame reactor. Furthermore, spray pyrolysis employing furnace reactor offers more quality at very less operational cost and least operative skill than flame reactors which have very high operating budgets, and the value of the resulting product in neither stoichiometry nor crystalline morphology leaves a proportion to be anticipated. The tubular furnace reactor that is used in the PSP offers a well-controlled temperature (heat) zone over prolonged period of time for conversion of the precursor to the required final product. In addition, tubular reactor method also provides the possibility for industrial scaling. Hence spray pyrolysis method is referred as a method that draws on the advantage of outstanding three-dimensional mixing of the reactants and was also established for multi component systems. These facts are generally one of the drivers of the high-quality tubular reactor used in this work as revealed here in **Figure 4**. **Figure 4** displays the split tube furnace and the aluminum tube that were employed in this study for the synthesis of ZnO NPs, respectively.

5. Pneumatic spray pyrolysis system

The spray pyrolysis system consists of a (i) split tube furnace, (ii) aluminum reactor as the reaction zone, (iii) nebulizer that converts the starting solution into droplets, (iv) sample holder (i.e. filter, electrostatic precipitator and thermophoretic sampler) and (v) an exhaust. The pneumatic spray pyrolysis (PSP) system was designed and assembled specifically for the fabrication of ZnO nanomaterials for solar cell applications. **Figure 5** displays the schematic diagram of the PSP technique system used in this study for the fabrication of ZnO nanostructures.

Pneumatic spray pyrolysis deposition involves forcing the precursor solution into fine nozzles to yield the mist of droplets of the precursor liquid at audible sound. The generated precursor droplets in a chamber are then transported into the furnace (heated zone) through the aluminum tube onto a preheated glass substrate. The PSP reaction was carried out at room temperature unlike in ultrasonic spray pyrolysis (USP) where the reaction is carried out in an oxygen-free environment using nitrogen or argon as a carrier gas to elude oxide formation and to declare reduction to metal that occurs in the high-temperature reaction zone. The spray

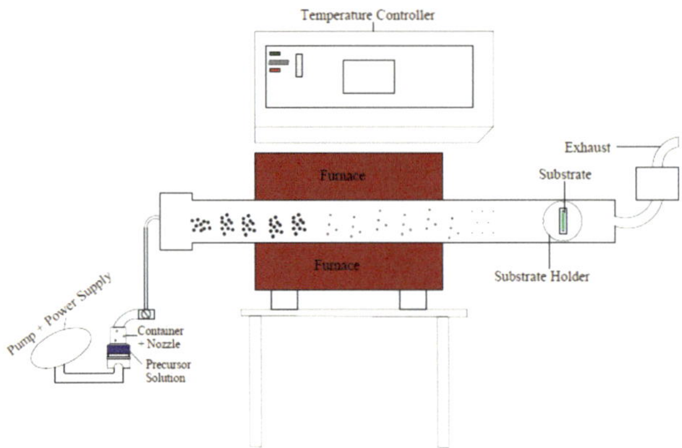

Figure 5.
Schematic diagram of the PSP system used for deposition of ZnO nanomaterials.

pyrolysis reaction was run at a constant flow rate that was automatically controlled by the pneumatic pump at an average flow rate of 5–6 ml/min. The spray pyrolysis deposition was accomplished at the system functioning at a temperature of 400°C; this is due to sufficient residence time consumed by spray vapor droplets inside the furnace. The deposition temperature and flow rate have the influence on the shape, size and the structure of the deposited thin film. The ZnO thin films were deposited on top of the fluorine-doped tin oxide (FTO) glass substrates.

5.1 Pneumatic spray pyrolysis mechanism for droplet generation

In order for the droplet to be generated, the precursor solution must have sufficient high velocity of ejection in the pneumatic vessel. The speed of the air from the pneumatic pump measures the droplet size in the pneumatic atomization. Hence increasing the pneumatic pump speed leads to the formation of very small droplets. The splitting of the precursor solution occurs to produce droplets and is transferred from the liquid pneumatic source interface to the surrounding air as a mist of very fine dense droplets. This process usually happens when the precursor solution is subjected to sufficiently high intensity of pneumatic field. Spraying of the liquid is done in the presence of very high pneumatic pump speed to the precursor solution which results in the formation of aerosols with constant droplet size and depends closely on the characteristics of the liquid. There are several main features which demonstrate scale pneumatic spray pyrolysis such as aerosol pneumatic generator, high temperature furnace with a fire wall heated reactor, electrostatic filter as well as the vacuum system. The precursor solution sheet can be interrupted to form droplets when the high-velocity air conveys its energy to the precursor. **Figure 6** shows the schematic diagram for the droplet generation at the crest of the waves.

Ultrafine droplets with good sphericity and uniform size distribution can be attained only if the energy of precursor sheet fragmentation can be delivered by the use of pneumatic energy [33]. The spray of very small ultra-fine droplets is properly produced by high pneumatic atomization speed. Traditional mechanical approaches such as pressure or gas-assisted methods are not considered to be advantageous on the generation of very small droplets compared to pneumatic atomization because it has some specific characteristics that make it more valuable [29, 34, 35]. Pneumatic atomization is a very productive method of generating small droplets. The ZnO nanostructures were formed after the precursor droplets have been generated.

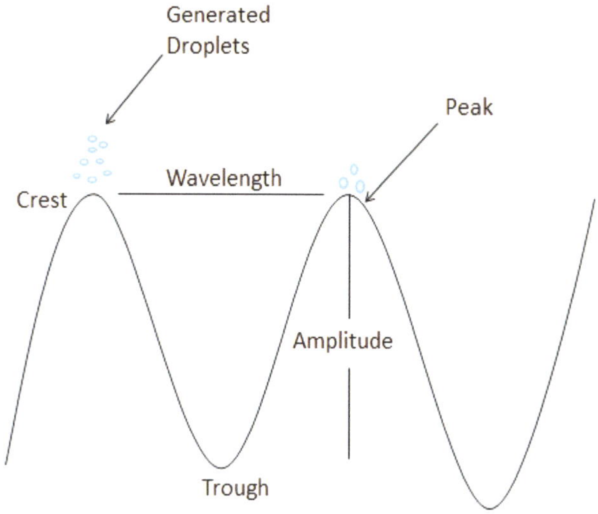

Figure 6.
Schematic presentation of droplet generation at the crest of capillary waves.

5.2 Pyrolysis of ZnO precursor solution

Generally, when a droplet hits the surface of the substrate, many processes take place simultaneously such as the vaporization of residual solvent, droplet diffusion as well as the corrosion of salt. Several design cores exist for the pyrolysis of the precursor depending upon the chemical environment. The mechanism of the reaction of zinc acetate aerosol for the formation of ZnO depends on the size of the droplet. The decomposition conditions are assumed to be similar to a CVD process if the majority of the aerosol is in gas phase when in contact with the substrate. The majority of aerosols with larger droplet sizes could not have enough time to vaporize completely, while those with small droplet sizes are easily decomposed by pyrolysis before reaching the substrate. Hence the droplet size distribution is small enough in the pneumatic spraying. At very low temperatures, the decomposition rate is assumed to be slower than the deposition rate which results in the formation of liquid film on the surface as revealed by process A in the above diagram. The resulting layer is slowly dried; however, it can still have several organics and possible cracks on it. At this stage a small amount of zinc oxide as a hydrated white precipitate will be present in its amorphous phase. **Figure** 7 shows the deposition processes of the aerosol droplet transport that occur with rising the temperature of the substrate.

As the substrate temperature increases to higher temperatures in process B, the solvent from the precursor solution vaporizes completely through the trip of the droplet just before striking the floor (surface), and the precipitate hits the substrate where corrosion of the aerosol takes place [36]. As the temperatures increase in process C, the solvent also evaporates before the droplets strike the substrate, and the solid precipitate melts and sublimes which results in diffusion of the vapor to the substrate and commences a chemical reaction process. This is the stage where the adherent films can mostly be achieved by CVD [36, 37]. At higher temperatures in process D, the precursor solution evaporates just before hitting the glass substrate; at this stage the vapor phase first experiences a chemical reaction before invading on top of the substrate which results in the formation of solid nanoparticles that remain to the floor of the substrate. Hence we can speculate that the ZnO solid films can be yielded by evaporation and corrosion of the precursor prior to striking the substrate at high temperatures. Additionally, the existence of huge nanoparticles on the surface

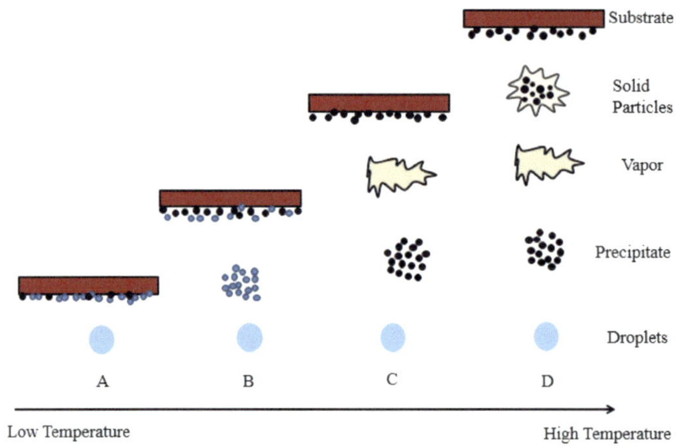

Figure 7.
Illustration of spray droplet transportation and deposition on the substrate at different temperatures.

of the substrate can be due to very large droplets, which are not completely decomposed when arrived at the substrate. The chemical reaction mechanism for the formation of ZnO nanoparticles is more similar to that proposed by Livage [35, 38] in the production of vanadium oxide NPs and similar to the one adopted by Taziwa et al. [29, 30] in the production of carbon-doped titanium dioxide NPs and is shown by the following reactions:

$$Zn(O_2CCH_3)_{2(s)} + 4CH_3CH_2OH_{(aq)} \rightarrow Zn(OCH_2CH_3)_{4(aq)} + CH_3CO_2H_{(aq)} \quad (1)$$

$$Zn(OCH_2CH_3)_{4(aq)} + CH_3CO_2H_{(aq)} \xrightarrow{TBA+Pyrolysis@400°C} ZnO_{2-x}C_{x(s)} + CO_{2(g)} + H_2 \quad (2)$$

6. Operation of PSP system

Pneumatic spray pyrolysis can be used to generate an aerosol from a dilute aqueous salt solution (ZnO precursor solution), resulting in the formation of nanoparticles with a narrow size distribution. The reaction was carried at a constant flow rate and was automatically controlled by the pneumatic pump which was used to blow the precursor solution in the PSP vessel. The pneumatic spray deposition was performed at the system operating at an average flow rate of 5–6 ml/min at a temperature of 400°C; this is probably due to sufficient residence times consumed by spray vapor droplets into the furnace. **Table 2** reveals the typical deposition factors of the chosen precursor.

Process parameters	Pneumatic spray pyrolysis @400°C
PSP pump air flow rate	5–6 ml/min
PSP vessel pressure (atm)	1
Precursor temperature (°C)	25
Precursor pH	1.27
Substrate type and temperature (°C)	Glass/400°C
Deposition time (h)	4
Deposition angle (°)	90
Annealing temperature (°C)	400

Table 2.
Operational conditions for PSP method.

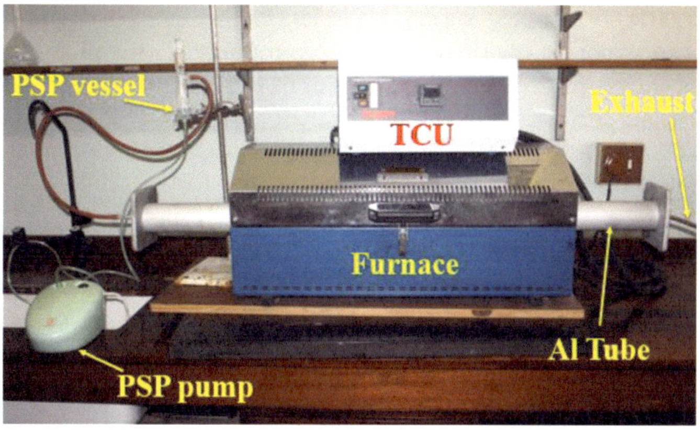

Figure 8.
Actual photograph of the horizontal pneumatic spray pyrolysis system utilized in the study in its final form.

1. Prepare the desired precursor solution as outlined in table 2 above in a 250 ml volumetric flask

2. Transfer the precursor solution into a Pneumatic spray pyrolysis vessel. Ensure that the PSP vessel is completely closed to avoid leakage of the precursor solution.

3. Into an Aluminum tubing/quartz tubing insert the aluminum substrate holder housing the F: SnO_2 glass substrate. Set the furnace to the desired temperature. Allow it to stand for 10-15 minutes. Until the substrate has reached the furnace temperature

4. Connect the inlet of the aluminum/quartz tubing to the outlet of the Pneumatic vessel and the outlet to the (1) exhaust for thin film synthesis and/or (2) to the nano powder collection system

5. Switch on the PSP pump to the desired flow rate and then carry out spray deposition.

6. To finish pneumatic spray deposition, simultaneously switch off the furnace and the PSP pump. Anneal the sample until they are ready for collection

Figure 9.
A schematic indicating the necessary steps to set up and perform pneumatic spray depositions for synthesis of pure ZnO and carbon-doped ZnO.

The pneumatic spray pyrolysis system in its final form is clearly revealed here in **Figure 8**; it is incorporated with all the other components as well as those listed in other sections above. In addition, the diagram in **Figure 9** highlights the necessary steps to set up and perform depositions of ZnO employing PSP system using a horizontal reactor. The aluminum substrate holder was employed in all conditions for ZnO thin film production.

7. Results and discussion

7.1 Scanning electron microscopy (SEM)

Figure 10 shows the SEM micrographs of the un-doped and C-ZnO NPs fabricated using the PSP system with different levels of carbon dopants. The SEM images of synthesized un-doped and C-ZnO samples have revealed the formation of spherical-shaped ZnO NPs with uneven grain size distribution. The SEM micrographs of the synthesized samples have revealed the change in morphology and shape of un-doped and C-ZnO NPs as the level of dopant increases. Additionally, the SEM images have revealed that the un-doped and C-ZnO samples consist of mesoporous morphology with a multiple porous network structure. Moreover, the introduction of carbon in the ZnO matrix has resulted in the formation of large spherical NPs surrounded by small NPs.

The inserts in **Figure 10** show an elemental analysis of the un-doped and C-ZnO NPs which was performed with energy dispersive X-ray (EDX) spectroscopy, using a scanning electron microscope (SEM). The EDX spectra of both un-doped and carbon-doped ZnO NPs revealed the existence of Zn, O and C which indicates the successful pyrolysis of zinc ethoxide to form ZnO nanomaterials. Additionally, the presence of the C, Zn and O elements in the carbon-doped samples indicates the

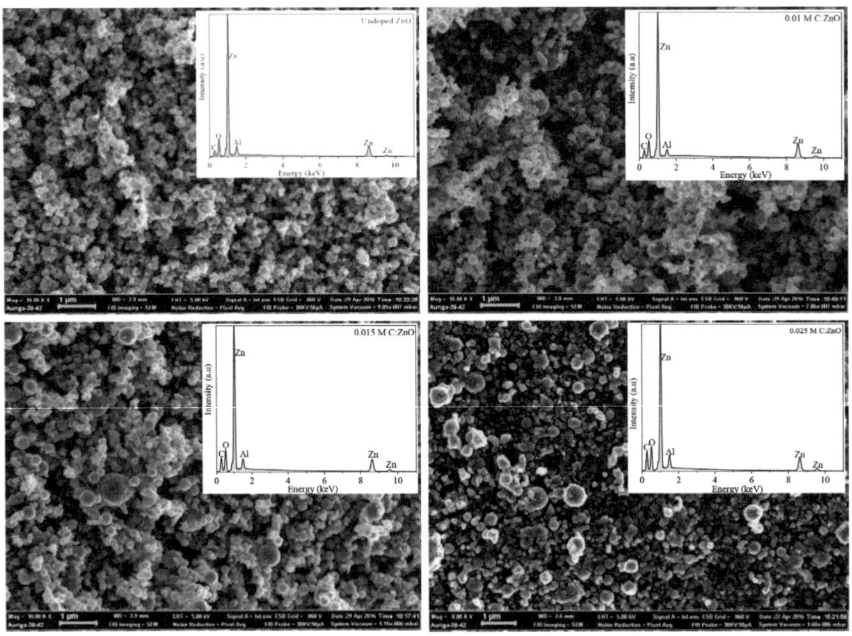

Figure 10.
SEM micrographs of (a) un-doped ZnO, (b) 0.01 M C-ZnO, (c) 0.015 M C:ZnO and (d) 0.025 M C:ZnO samples. Moreover, inserts in (a), (b), (c) and (d) show typical EDX spectra of the synthesized un-doped and C-ZnO nanostructures.

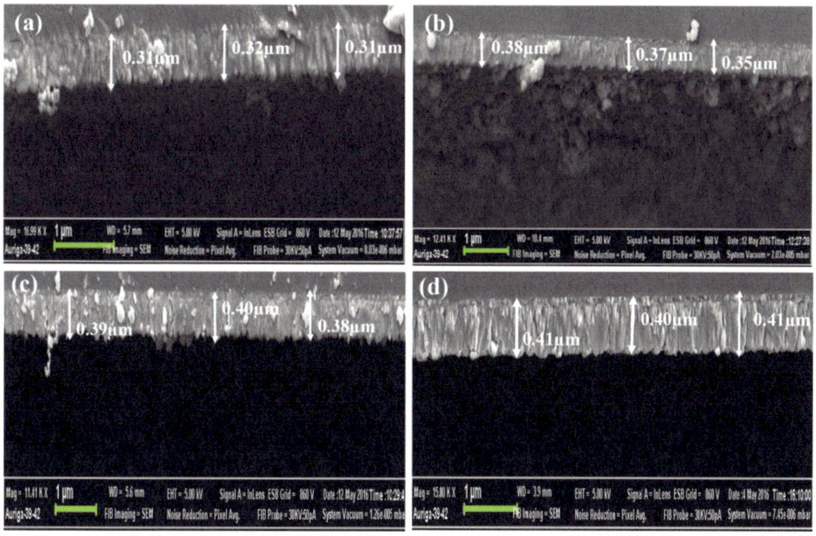

Figure 11.
The cross-sectional SEM micrographs of (a) un-doped ZnO, (b) 0.01 M C-ZnO, (c) 0.015 M C:ZnO and (d) 0.025 M C:ZnO samples

effective modification of carbon in the ZnO matrix. Furthermore, the EDX spectra of the synthesized ZnO NPs have revealed the existence of Al in all the samples; this is due to the aluminum tube reactors and aluminum substrate holder used in the pneumatic spray pyrolysis system.

7.2 Cross-sectional SEM

Figure 11 shows cross-sectional SEM images of the un-doped and carbon-doped ZnO thin films deposited by pneumatic spray pyrolysis technique. The thickness of the thin films was measured using cross-sectional SEM method. It is observed that the film thickness changes with increasing dopant concentration from 0.31 to 0.41 μm. The thickness of the un-doped ZnO thin film is 3.1–3.2 which was gradually increased due to the presence of dopant to 0.3538 at lower concentrations of dopant and 0.38–0.41 at higher concentrations. This implies that the introduction of carbon to ZnO lattice also affects the film thickness of the fabricated ZnO NPs as evidenced by **Figure 11**.

7.3 X-ray diffraction

Figure 12 shows the X-ray diffraction patterns of the ZnO NPs fabricated by PSP technique and recorded in the 2θ range of 30–80° at room temperature. The XRD patterns of both un-doped and C-ZnO samples have displayed the characteristic peaks of the hexagonal Wurtzite structure. The diffractograms obtained at room temperature for both un-doped and C-ZnO were observed by XRD lines at 31.90, 34.50, 36.34, 47.73, 56.88, 63.04, 68.20 and 77.33°. These lines are indexed as (100), (002), (101), (102), (110), (103), (200) and (112), respectively. The sharp and intense peaks for the dominant peak at 36.34° indexed (101) in the XRD diffractograms show that the synthesized samples are highly crystalline.

The effect of carbon doping on the ZnO lattice was cross-examined by monitoring the three dominant peak positions of (100), (002) and (001) planes. The introduction of carbon doping resulted in a peak shift to higher 2θ wave numbers,

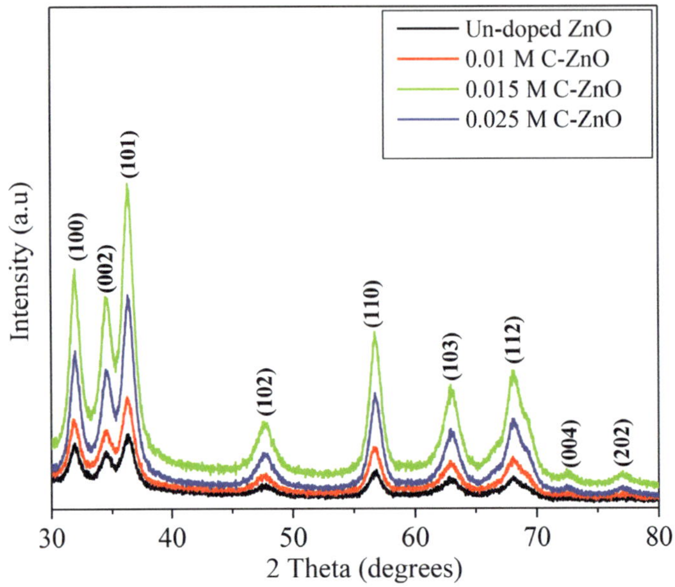

Figure 12.
XRD patterns of ZnO NPs synthesized by PSP technique.

which indicates substitutional doping in ZnO samples. In this work, the crystallite size was calculated using the Scherrer method which is considered as a standard method. The Scherrer method has revealed that the crystallite size increases as the dopant levels increase. The calculated crystallite sizes for both un-doped and C-ZnO were 9.60, 9.99, 0.96 and 10.22 nm for un-doped ZnO, 0.01 M C-ZnO, 0.015 M C:ZnO and 0.025 M C:ZnO samples, respectively.

7.4 Diffuse reflectance spectroscopy (DRS)

Figure 13 shows the UV-Vis diffuse reflectance spectra (DRS) of the synthesized un-doped and C-doped ZnO NPs. The UV-Vis DRS analysis has shown that there is a shift in absorption edge as the dopant level increases. The energy band gaps of the ZnO NPs were estimated by using Eg = 1239/λEdge*eV. The absorption

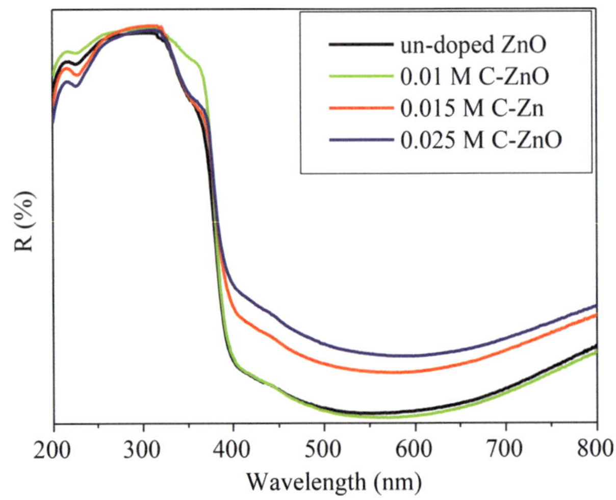

Figure 13.
The DRS spectra of the unmodified and C-ZnO samples synthesized by PSP system.

edge of the un-doped ZnO sample is 374 nm with an energy band gap 3.31 eV which is larger than for bulk ZnO NP with an absorption edge at 388 nm and energy band gap of 3.2 eV as revealed by the UV-Vis analysis. The absorption edges of the carbon-doped samples have revealed a red shift with energy band gaps of 3.29, 3.28 and 3.27 eV for the 0.01, 0.015 and 0.025 M of C-ZnO samples, respectively. This red shift in energy band gaps is probably due to size confinement effect.

8. Conclusion

Gathering knowledge about the challenges that other researchers experienced when working with spray pyrolysis for the production of ZnO together with understanding the properties and the crystal structure of ZnO has made it possible to design a novel pneumatic spray pyrolysis (PSP) system from ultrasonic spray pyrolysis (USP) for the deposition of ZnO NPs. The novel PSP system developed has presented unique features in material synthesis of ZnO nanostructures, like using the horizontal furnace reactor as compared to the vertical systems used in other techniques such as CVD, sol-gel, ion-assisted deposition etc. The horizontal system offers several advantages for thin film deposition in the absence of any tailing effect detected in the partial or oblique angle illustrations like in almost all the CVD methods. The samples in this system were deposited at 90° angle, which enables the aerosol beam comprised of the precursor solution vapor to directly cooperate with the substrate consistently. This system was able to produce the desired ZnO nanostructured properties for solar cell application. The SEM micrographs of both un-doped and carbon-doped samples have revealed the formation of spherical-shaped ZnO nanoparticles mesoporous morphology. The SEM images also revealed that the morphology and shape of the fabricated ZnO samples change as the dopant level increases. Additionally, the EDX analysis has confirmed the presence of Zn, C and O in the synthesized samples which indicates the successful pyrolysis of zinc ethoxide solution to form ZnO nanoparticles. The cross-sectional SEM has revealed the increase in film thickness as the dopant levels increase from 0.31 to 0.41 μm. The XRD has revealed the characteristic peaks of the hexagonal Wurtzite structure of ZnO for both un-doped and carbon-doped ZnO samples. XRD lines were observed at 31.90, 34.50, 36.34, 47.73, 56.88, 63.04, 68.20 and 77.33° and were indexed as (100), (002), (101), (102), (110), (103), (200) and (112), respectively. Additionally, the XRD analysis has also revealed a shift in the peaks to higher 2θ standards, which indicates the substitutional doping in the synthesized carbon-doped samples. Lastly, the UV-Vis DRS analysis has revealed a blue shift in absorption spectra of the synthesized samples with an increase in carbon doping.

Acknowledgements

The authors are grateful to the financial funding from their sponsors, the South African National Research Foundation (NRF), the Govan Mbeki Research and Development Centre (GMRDC) of the University of Fort Hare and the Sasol Inzalo Foundation. The authors would also like to acknowledge the DST/CSIR Nanotechnology Innovation Centre, National Centre for Nanostructured Materials, CSIR, and Centre of Image analysis and the University of Cape Town for the characterization of the ZnO NPs.

Author details

Luyolo Ntozakhe[1*] and Raymond Tichaona Taziwa[2]

1 Fort Hare Institute of Technology (FHIT), University of Fort Hare, Republic of South Africa

2 Department of Applied Science, Walter Sisulu University (Postdam Campus), East London, Republic of South Africa

*Address all correspondence to: lntozakhe@ufh.ac.za

References

[1] Galstyan V, Comini E, Baratto C, Ponzoni A, Faglia G, Bontempi E. Growth and Gas Sensing Properties of Self-Assembled Chain-Like ZnO Nanostructures. Process Engineering. 2012;**47**:762-765

[2] Banerjee AN. The design, fabrication, and photocatalytic utility of nanostructured semiconductors: Focus on TiO2-based nanostructures. Nanotechnology Science and Applications. 2011;**4**:35-65

[3] Zhou Q, Wen JZ, Zhao P, Anderson WA. Synthesis of Vertically-Aligned Zinc Oxide Nanowires and Their Application as a Photocatalyst. Journal of Nanomaterials. 2017;**7**:1-17

[4] Zhang Y, Ram MK, Stefanakos EK, Goswami DY. Synthesis, Characterization, and Applications of ZnO Nanowires. Journal of Nanomaterials. 2012;**2012**:1-22

[5] Cavallo C, Pascasio FD, Latini A, Bonomo M, Dini D. Nanostructured Semiconductor Materials for Dye-Sensitized Solar Cells. Journal of Nanomaterials. 2017;**2017**:1-31

[6] Bhakat C, Singh PP. Zinc Oxide Nanorods: Synthesis and Its Applications in Solar Cell. International Journal of Modern Engineering Research. 2012;**2**:2452-2454

[7] Aisah N, Gustiono D, Fauzia V, Sugihartono I, Nuryadi R. Synthesis and Enhanced Photocatalytic Activity of Ce-Doped Zinc Oxide Nanorods by Hydrothermal Method. Materials Science and Engineering. 2017;**172**:899-1757

[8] Badreddine K, Kazah I, Rekaby M, Awad R. Structural, Morphological, Optical, and Room Temperature Magnetic Characterization on Pure and Sm-Doped ZnO Nanoparticles. Journal of Nanomaterials. 2018;**2018**:1-11

[9] Zhang X, Qin J, Hao R, Wang L, Shen X, Yu R, et al. Carbon-Doped ZnO Nanostructures: Facile Synthesis and Visible Light Photocatalytic Applications. Journal of Physical Chemistry C. 2015;**119**:20544-20554

[10] Nirmala M, Anukaliani A. Synthesis and characterization of undoped and TM (Co, Mn) doped ZnO nanoparticles. Materials Letters. 2011;**65**:2645-2648

[11] Ozturk S, Kilinic N, Tsaltin N, Ozturk ZZ. Fabrication of ZnO nanowires and nanorods. Physica E, Low-Dimensional Systems & Nanostructures. 2012;**44**:1062-1065

[12] Mah CF, Beh KP, Yam FK, Hassan Z. Rapid formation and evolution of anodized zn nanostructures in NaHCO3 solution. ECS Journal of Solid State Science and Technology. 2016;**5**:105-112

[13] Davinder R, Mukesh K, Sandeep KA. Deposition of nanocrystalline thin TiO2 films for MOS capacitors using Sol-Gel spin method with Pt and Al top electrodes. Solid State Electronics. 2012; **76**:71-76

[14] Zaleta-Alejandrea E, Camargo-Martineza J, Ramirez-Gariboa A, Pérez-Arrietab ML, Balderas-Xicohténcatla R, Rivera-Alvareza Z, et al. Structural, electrical and optical properties of indiumchloride doped ZnO films synthesized by Ultrasonic Spray Pyrolysis technique. Thin Solid Films. 2012;**524**:44-49

[15] Yoo Y, Bruckenstein S. Synthesizing nanoparticles using reactions occurring in aerosol phases. Advances in Nanoparticles. 2013;**2**:313-317

[16] Barthwal S, Kim YS, Lim SH. Fabrication of amphiphobic surface by using titanium anodization for largearea three-dimensional substrates. Journal of Colloid and Interface Science. 2013;**400**: 123-129

[17] Jyotia M, Vijayb D, Radhac S. To Study the role of temperature and dodium hydroxide concentration in the synthesis of Zinc. International Journal of Scientific Research Publications. 2013;**3**:1-11

[18] Bendavid A, Martin PJ, Preston EW. The effect of pulsed direct current substrate bias on the properties of titanium dioxide thin films deposited by filtered cathodic vacuum arc deposition. Thin Solid Films. 2008;**517**:494-499

[19] Anders A, Lim S, Yu KM. High quality ZnO:Al transparent conducting oxide films synthesized by pulsed filtered cathodic arc deposition. Thin Solid Films. 2010;**518**:3313-3319

[20] Baji Z, Labadia Z, Molnara G, Pecza B, Vad K, Horvatha ZE, et al. Highly conductive epitaxial ZnO layers deposited by atomic layer deposition. Thin Solid Films. 2014;**562**:485-489

[21] Chun H, Takeshi S, Hiroyuki U, Yoshiki S, Naoto K. Fabrication of ZnO nanoparticles by pulsed laser ablation in aqueous media and pH-dependent particle size: An approach to study the mechanism of enhanced green photoluminescence. Journal of Photochemistry and Photobiology A: Chemistry. 2007;**191**:66-73

[22] Foo KL, Kashif M, Hashim U, Ali ME. Fabrication and Characterization of ZnO Thin Films by Sol-Gel Spin Coating Method for the Determination of Phosphate Buffer Saline Concentration. Current Nanoscience. 2013;**9**

[23] Kawabata K, Nanai Y, Kimura S, Okuno T. Fabrication of ZnO nanoparticles by laser ablation of sintered ZnO in aqueous solution. Applied Physics A. 2012;**107**:6745

[24] Yuan X, Xua W, Huanga F, Cheng D, Wei Q. Polyester fabric coated with Ag/ZnO composite film by magnetron sputtering. Applied Surface Science. 2016;**390**:863

[25] Kaushika VK, Mukherjeeb C, Gangulic T, Sena PK. Material characterizations of Al:ZnO thin films grown by aerosol assisted chemical vapour deposition. Journal of Alloys and Compounds. 2016;**689**:1028

[26] Raghu P, Srinatha N, Naveen CS, Mahesh HM, Angadi B. Investigation on the effect of Al concentration on the structural, optical and electrical properties of spin coated Al:ZnO thin films. Journal of Alloys and Compounds. 2017;**68**:694

[27] Rahman MA, Phillips MR. Ton-That C. Efficient multi-coloured Li-doped ZnO thin films fabricated by spray pyrolysis. C. Journal of Alloys and Compounds. 2017;**691**:339

[28] Bagabas A, Alshammari A, Aboud MF, Kosslick H. Room-temperature synthesis of zinc oxide nanoparticles in different media and their application in cyanide photo-degradation. Nanoscale Research Letters. 2013;**8**:516

[29] Mwakikunga BW. Progress in ultrasonic spray pyrolysis for condensed matter sciences developed from ultrasonic nebulization theories since michael faraday. Critical Reviews in Solid State and Materials Sciences. 2014;**39**:46-80

[30] Taziwa R, Meyer E. Carbon Doped Nano-Crystalline TiO 2 Photo-Active Thin Film for Solid State Photochemical Solar Cells. Advances in Nanoparticles. 2014;**3**:54-63

[31] Summary Related to Spray Freeze Dried Microspheres by means of an Ultrasonic Spray Nozzle. https://microspray.com/spray-freeze-dried-microspheres-ultrasonic-spray-nozzle-7268112/. [Accessed, 26 September 2018]

[32] Ali A, Zafar H, Zia M, Haq I, Phull AR, Ali JS, et al. Synthesis,

characterization, applications, and challenges of iron oxide nanoparticles. Nanotechnology. Science and Applications. 2016;**9**:49-67

[33] Rajan R, Pandit AB. Correlations to predict droplet size in ultrasonic atomization. Ultrasonics. 2001;**39**:235-255

[34] Kurosawa M, Futami A, Higuchi T. Characteristics of liquids atomization using surface acoustic wave. Mechanical Engineering. 1997;**2**:801-804

[35] Dalmoro A, d'Amore M, Barba AA. Droplet size prediction in the production of drug delivery microsystems by ultrasonic atomization. Translational Medicine @UniSa. 2013;7:6-11

[36] Perednis D, Gauckler LJ. Thin Film Deposition Using Spray Pyrolysis. Journal of Electroceramics. 2005;**14**:103-111

[37] MateiGhimbeua C, Van Landschoot RC, Schoonman J, Lumbreras M. Preparation and characterization of SnO2 and Cu-doped SnO2 thin films using electrostatic spray deposition (ESD). Journal of the European Ceramic Society. 2007;**27**:207-213

[38] Suh WH, Suslick KS. Magnetic and porous nanospheres from ultrasonic spray pyrolysis. Journal of the American Chemical Society. 2005;**127**:12007-12010

Chapter 6

Surface-Enhanced Raman Spectroscopy (SERS) Based on ZnO Nanorods for Biological Applications

Sanghwa Lee and Jun Ki Kim

Abstract

Detection of nanometer-sized biomarkers is a research topic that attracts much attention as an application for early diagnosis of diseases. Biopsy monitoring by analyzing cell secretion in a non-destructive way has many advantages in the field of biomedicine. We introduce the Raman signal enhancement method on a biosensing chip based on surface-enhanced Raman diagnosis. This approach has the advantage because the ZnO nanorods are grown to form nanoscale porosity and are coated with gold to enable size selective biomarker detection. After sputtering gold on the grown ZnO nanostructures, the unique feature of clustering the nanorod's heads first appeared. The grain formation on the head was the main factor for the localized surface plasmon resonance (LSPR) enhancement, and this fact could be verified by finite element analysis. It has been demonstrated in breast cancer cell line that the cell viability is also high in such gold-clad ZnO nanostructure-based surface-enhanced substrates. For bioapplication, interstitial cystitis/bladder pain syndrome (IC/BPS) animal model was prepared by injecting HCl into the bladder of a rat, and urine was collected a week later to conduct Raman spectroscopy experiments.

Keywords: surface-enhanced Raman spectroscopy (SERS), ZnO nanorods, gold clustering, finite element method (FEM), cell viability, interstitial cystitis/bladder pain syndrome (IC/BPS)

1. Introduction

Biological particles on the nanometer and submicrometer scale, such as proteins, lipids, nucleic acids, exosomes, and metabolic content, have attracted much attention as biomarkers for diagnosing diseases from biologically generated fluids such as blood, urine, and lymph. These biomarkers are now understood to be fundamental to healthy intercellular communication and can be produced in diseased cells. Label free Raman spectroscopy is useful for verification of biological samples ranging from nanoscale to millimeter size, such as tissue [1, 2], cells [3–5], bacteria [6, 7], exosomes [8, 9], and proteins [10, 11]. After incident laser emission with a single wavelength, Raman spectroscopy can identify biomarkers with the spectral peak position as a fingerprint because the molecular vibrations of the sample are represented by spectra due to inelastic scattering. Electromagnetic enhancement can be achieved on rough surface of metal such as a gold or silver nanoparticle that causes amplification of the light by local surface plasmon resonance (LSPR) effects [12, 13].

A "hot spot" is formed on the surface of the SERS particles, and the Raman signal is dramatically increased at the nano-sized gap. Surface-enhanced Raman spectroscopy (SERS) is an approach for cell analysis and identification that applies a wide range of chemical spectroscopy to nanometer-sized biomarkers. Recent studies on monomolecular scales have been made possible through surface-enhanced Raman techniques [14–18]. According to finite element method (FEM) analysis, when the colloid is separated by 2 nm between a diameter of 30 nm colloids, a "hot spot" is formed which gives a surface-enhanced effect of about 10^8 degrees [13]. In biomedical applications, biomarkers suitable for these nanogaps are very rare, and due to the size and shape of biomolecules, research on nanogap and signal enhancement of the SERS structure is needed to optimize the LSPR effect.

In this chapter, we fabricated SERS substrate based on ZnO nanorods and improved the SERS effect by forming selective growth clustering of gold nanoparticles, which could be formed in specific condition of ZnO nanorod-based SERS substrate. To control the porosity and gold nanostructure, the length and density of the ZnO nanostructures and the thickness of the deposited gold were modified morphologically. The SERS enhancement mechanism was described based on finite element analysis. Cell viability was also evaluated to determine the presence or absence of toxicity for cancer cell applications. In other bio-applications, we demonstrate early diagnostic possibilities with Raman signals and statistical analyses from nano-sized biomarkers of intractable inflammatory diseases that cause patient pain.

2. Nanorod manufacturing method

Most of the research to fabricate SERS-based chips focuses on optimizing the surface of substrates through nanomaterials and nanostructures synthesized using sophisticated techniques such as lithographic patterning or high-temperature processes [16, 19–22]. Other research groups deposit Au/Ag nanoparticles on papers [23–25] or coat metal on a Si nanowire structure [26] to make a porous SERS substrate suitable for biological or liquid samples. Such Si nanowires are too dependent on the substrate and are difficult to combine with common cell culture substrates such as glass and petri dishes, due to their amorphous and manufacturing nature. In the case of paper-based SERS substrate, porosity and nanogaps could not be adjusted. On the other hand, if a ZnO nanorod-based platform is introduced, the substrate can be manufactured at a temperature below 100°C. Furthermore, homogeneous nanostructures can be formed without any lithography process on amorphous substrates such as glass and plastic, which are common in bioscience applications [27, 28].

To make the SERS substrates, the Si wafer were scribed with a size of 1 × 1 cm^2 for substrate of ZnO nanorods initially. It was cleaned in ethanol and deionized (DI) water for 5 min, respectively. The 30 nm ZnO seed layer was deposited on the surface of as-prepared samples by using RF magnetron sputtering for 5 min under 100 W power to grow the vertically aligned ZnO nanorods utilizing by the hydrothermal synthesis. The ZnO growth solution was prepared by dissolving 10 mM zinc nitrate hexahydrate (Sigma-Aldrich Co.) and 0.9 mL of ammonium hydroxide (Sigma-Aldrich Co.) in 30 mL DI water. A homogeneous aqueous solution was achieved using mildly stirred vortexer for 5 min at room temperature. Then, the as-prepared samples were immersed into the aqueous solution in an oven at 90°C for 50 min.

3. SERS metal growth and enhancement test

After ZnO growth, the substrates were cleaned with DI water and dried with nitrogen gas. Finally, the ZnO nanorods (NRs) were coated with Au using a thermal evaporator (Alpha Plus Co., Ltd., Korea). The thickness monitor for 100 and 200 nm deposition was standardized. The morphological and structural properties of the Raman measured samples were observed by using a field-emission scanning electron microscope (FE-SEM) (S-4700, HITACHI, Japan) with 15 kV beam voltage. The procedure of the experiment including the measurement analysis is schematically shown as a diagram in **Figure 1**.

To obtain adequate porosity for the solution sample, the ZnO seed layer was modified and deposited such that the preferential growth direction of the zinc oxide nanorods was within about 10° from vertical. A volume of gold having a height of 100 and 200 nm per unit area was deposited on nanorods having length distribution of 300–450 nm or 500–650 nm, respectively. These four specimens were displayed with FE-SEM images of the 45° tilt view as shown in **Figure 2**. The top and bottom of the **Figure 2e** show the substrates with gold deposited (top) and not deposited (bottom) for **Figure 2a**, and ZnO is fully covered even when only 100 nm of gold is deposited. When the gold deposition is increased to 200 nm, the rod thickness is distributed about 10–30 nm thicker than when the gold deposition is not performed. Also, since the nanorod length distribution has a standard deviation of 50 nm and the deposited gold is clustered at the head of the nanorods, the height distribution of the gold clusters undergoes a similar variation. Therefore, when confocal Raman spectroscopy measurements are focused on the gold clusters, the head size can be a key factor in the Raman enhancement effect.

The Raman enhancement effect of SERS substrate based on ZnO nanorods was confirmed using 1 mM Rhodamine B drop, and the signals were measured after natural drying. Rhodamine B (RhB, >95%) purchased from Sigma-Aldrich was used as a standard for Raman measurements due to its refined condition. Raman measurements (LabRam Aramis, Horiba) were carried out using a 785 nm diode laser in a confocal geometry with a 0.5 NA, x50 objective lens and beam spot diameter ~1.9 μm. The spectrum of each point was measured in the range of 400–2500 cm^{-1} with a spectral resolution of 5 cm^{-1} and an integration time of 30 s at room

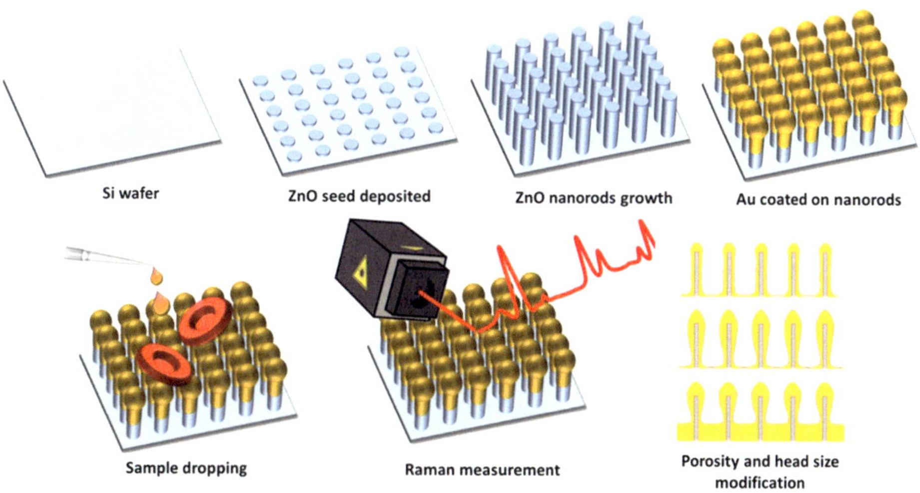

Figure 1.
Schematic of the experiment involving zinc oxide nanostructure-based SERS substrate fabrication.

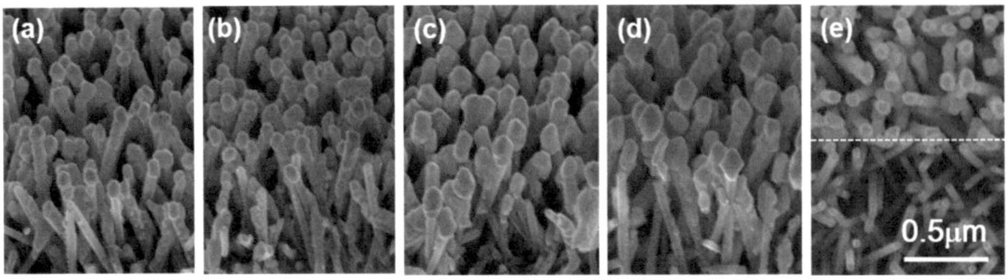

Figure 2.
Secondary electron images of the substrate with nanorod length and deposited gold thickness modified for (a) ZnO of length 400 nm with 100 nm deposited Au, (b) 600 nm length ZnO with 100 nm Au, (c) 400 nm length ZnO with 200 nm Au, and (d) 400 nm ZnO with 200 nm Au. (e) Shows the initial difference in covering due to gold coating. All of scales are the same.

temperature. The spectra were postprocessed by Savitzky-Golay smoothing, and a third-order polynomial fit to the autofluorescence background was subtracted.

Figure 3a shows the enhancement of the Raman signal ranging from 1000 to 1500 cm^{-1} according to each specimen. The greatest enhancement was observed in 600 nm ZnO nanorods deposited with 200 nm gold, which is a correction to the area of the circle by drying droplet. However, without area correction, a random point of 400 nm in length with the same average gold thickness appeared to give a greater improvement. This difference is explained by the correlation between signal enhancement and sample concentration. The enhancement factor EF follows the equation:

$$\mathbf{EF} = \left(\frac{I_{SERS}}{I_{bare}}\right)\left(\frac{C_{bare}}{C_{SERS}}\right)$$

where C_{SERS} is the concentration of RhB on the ZnO nanorod-coated Au SERS substrate, I_{SERS} is the measured Raman intensity from the nanorod-coated substrate, and C_{bare} and I_{bare} are the same quantities on the bare substrate, respectively. The effective concentration of RhB on the dried specimen varies across the sample as the diffusion of the droplet depends upon the porosity of the SERS substrate. In addition, even though samples of the same volume are dropped for all experiments, the value of C_{bare}/C_{SERS} is dependent upon a correction for area since the initial droplet varies in size. Therefore, EF due to the substrate correction can be seen to be largest in the 200 nm gold-coated nanorods of 600 nm length. This enhancement is independent for each specific peak, as shown in **Figure 3b**. This measurement shows that the enhancement effect is compared without denaturation of the sample, since the relative ratio between the peaks does not change significantly.

In addition, the SERS substrates based on zinc oxide nanorods show no coffee ring effect, as shown in **Figure 3c**. Due to the rise in concentration at the ring region, the edges of RhB on bare substrates and of RhB on gold thin films show stronger Raman signals. On the other hand, nanorod substrates have larger values in the interior of the deposition ring as shown in **Figure 3d**.

4. Localized surface plasmon resonance (LSPR) analysis

Calculation analysis using FEM was performed to understand the effect of Au cluster size on SERS enhancement. The finite element method (FEM) was used in COMSOL Multiphysics software (COMSOL Inc., USA) to simulate the SERS activities of the electromagnetic fields. A two-dimensional model for metal-coated

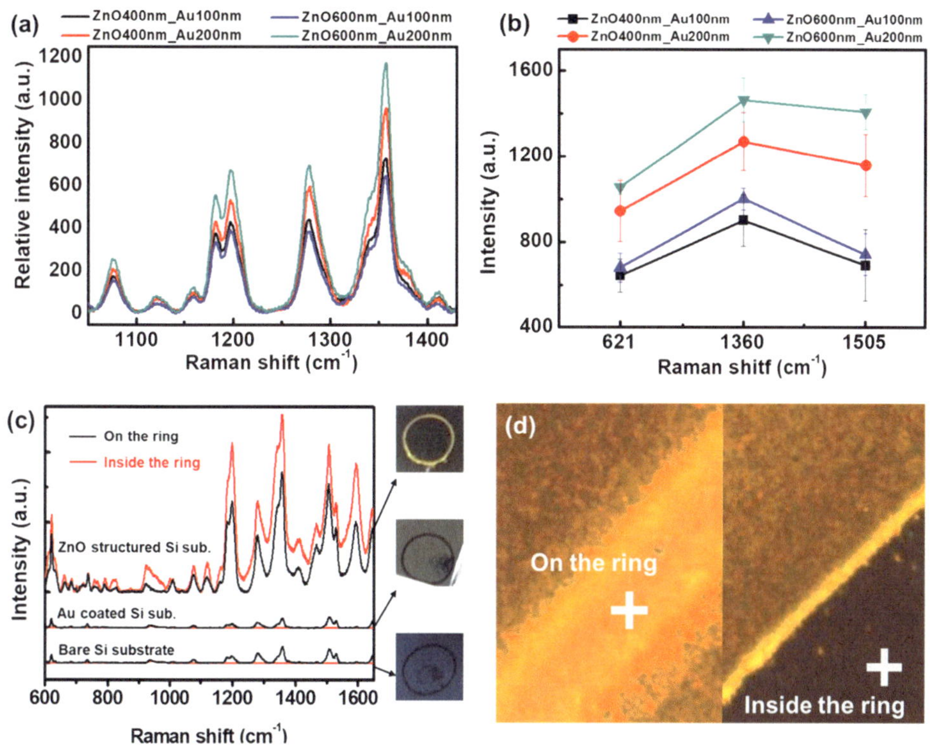

Figure 3.
(a) Raman signal enhancement of RhB on each substrate. (b) Differences in the enhancement intensity of each specific peak according to the substrate. (c) Differences in the coffee ring effect of the Raman signal depending on the nanostructure. (d) Optical microscope image including Raman acquired point.

nanostructured substrates with various metal spherical diameters has been established. The near-field distribution of the electromagnetic field was calculated for given boundary condition to solve the equation of time harmonic Maxwell at the excitation wavelength of 785 nm.

From secondary electron microscopy images, structures were modeled with 80 and 125 nm heads on nanorods of width 50 nm including gold coating and length

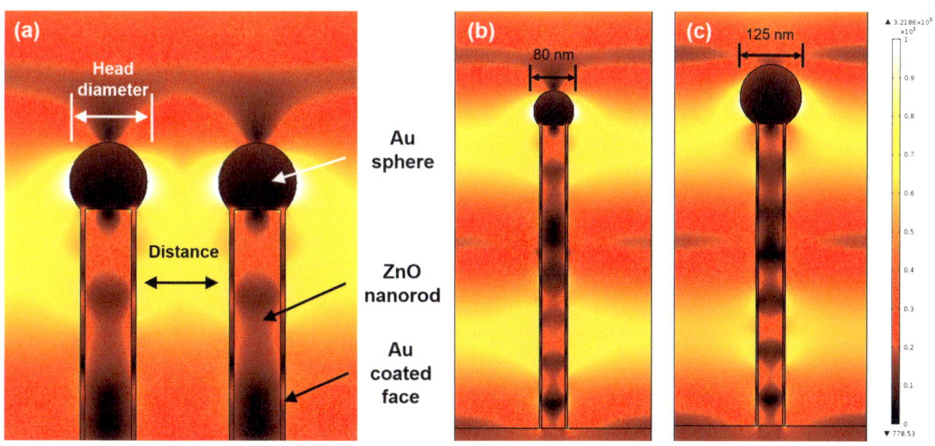

Figure 4.
Finite element analysis showing (a) the different factor in LSPR on ZnO nanorod-based SERS substrates with gold head diameters of (b) 80 nm and (c) 125 nm, respectively.

600 nm. The near-field distribution of the electric fields was calculated for 785 nm incident light and parallel plate boundary conditions with symmetry of the electric field. Even if the head width is increased, there is almost no change in the full scale, and the electric field is distributed in the vertical direction of the incident light as shown in **Figure 4**. The cross-sectional width of incident light is 2 μm, and the density of the nanorods in the region is constant. This suggests that the difference in SERS enhancement by Au thickness is due to concentration of sample at LSPR area.

5. Cell viability test

Raman measurements of nanometer-sized biomarkers secreted from living cells require confirmation of the suitability of cells for SERS substrates. The toxic endogenous properties of gold nanoparticles have previously been reported [29], and ZnO nanorods are reported to be toxic to NIH 3 T3 fibroblasts [30]. Therefore, it is necessary to confirm whether the sensing chip is suitable for cell application through the evaluation of cytotoxicity, and cell culture and cell viability tests were carried out as follows.

Breast cancer cell line of MDA-MB-231 was purchased from the Korean Cell Line Bank (Seoul, Korea). The breast cancer cells MDA-MB-231 were cultured in Dulbecco's modified Eagle's minimal essential medium (DMEM; Life Technologies, Inc., Grand Island, NY, USA) supplemented with 10% fetal bovine serum (FBS; Hyclone Laboratories, Logan, UT, USA) and a 1% penicillin–streptomycin solution (Life Technologies, Inc.) in a humidified 5% CO2 incubator at 37°C. Cell viability analyses were based on MTT (Sigma, USA) assays. Cells were plated at a density of 3 × 105 cells/well and incubated for 24 hours. After preparing the Au-ZnO substrate, the cell was treated with 5 mg/ml MTT for 30 min and then dissolved using DMSO. The absorbance was measured at 540 nm with an ELISA microplate reader (Multiskan EX, Thermo Scientific, USA). **Figure 5** shows the results of MTT assay of the death of the breast cancer cell line MDA-MB-231 according to SERS substrate condition. The MDA-MB-231 is commonly used in biotechnology applications such

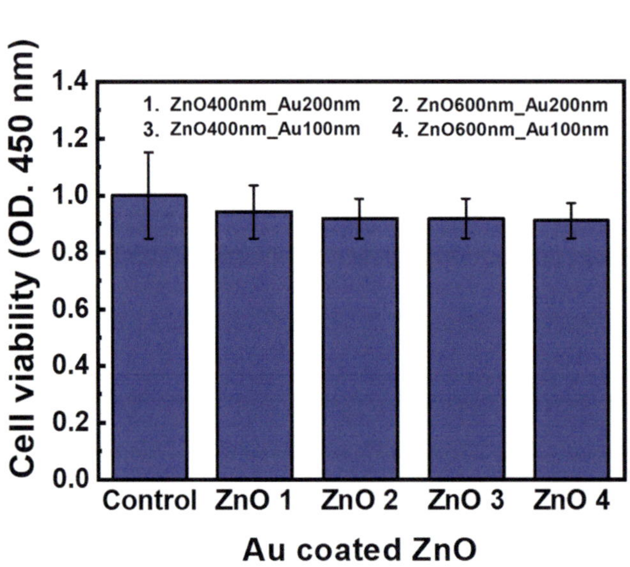

Figure 5.
Cell viability of MDA-MB-231 on each substrate by MTT assay.

as analyzing cancer cell malignancy. From the results, the substrate at this ZnO nanorod and gold deposition conditions is suitable for experiments in live breast cancer cell lines.

6. Bio-application

Interstitial cystitis/bladder pain syndrome (IC/BPS) is a refractory disease that afflicts the vague pelvic pain when the urine enters the bladder and makes frequent urination [31]. There are various treatments based on oral agents [32–34], but they are still unsatisfactory with frequent recurrence of symptoms and Hunner's lesions [35]. Parallel with the development of therapeutic technology, early diagnosis of IC/BPS related to quality of life and detection of disease before development to chronic type can minimize patient's pain and increase treatment effect. Therefore, it is necessary to confirm the possibility of early detection of nanometer biomarkers from urine obtained from interstitial cystitis animal model using SERS substrate. IC/BPS animal models and comparative groups for these experiments were derived using 10-week-old female Sprague Dawley rats. The rats were instilled with 0.2 M HCl for 10 min using a 26-gauge angiocatheter in the bladders of four rats, followed by neutralization and saline wash. Other rats in the comparison group were used as vehicle instead of HCl injection. The voiding pattern was analyzed to confirm the reproducibility of animal modeling as in the previous paper [36]. Rat urine was collected in a 50 mL tube using a metabolic cage. The voiding pattern measured at a week after HCl injection was examined, and the collected urine was used as a sample for Raman measurement. Twenty-four hours of natural voiding patterns in the metabolic cage were recorded and analyzed using AcqKnowledge 3.8.1 software and an MP150 data acquisition system (Biopac Systems, Goleta, CA, USA) at a sampling rate of 50 Hz. The volume change of the obtained raw data of urine was estimated by 1 mL unit as shown in **Figure 6a**. Irregular frequency of urinary dysfunction caused by bladder inflammation is observed in HCl-treated rats and is consistent with previous animal model studies [35, 36]. Steps and terraces in the graph are the excretion urine and the duration between voiding, respectively. The total amount of control and IC/BPS animal models for approximately 10 hours is 11 and 13 mL, respectively. However, compared with the amount, the frequency is 3 and 6 times. When the unit of step is 0.5 mL, the frequency is 4 and 11 times, and the difference in the voiding frequency is clearly revealed.

From the identified sample, a drop of 5 μL was applied to the SERS specimen, and the sample was allowed to spread for 60 min. After confirming that the droplets were dry and diffused, they were loaded onto a Raman spectroscope system attached to a microscope (IX-73, Olympus, Japan) and measured. As shown in **Figure 6b**, the diffused region of the sample can be confirmed by an optical microscope, and the region where the sample is diffused as in (c) can be confirmed to have a nanometer-scale porosity. In this area, Raman spectra were collected using a customized spectrometer (FEX-INV, NOST, Korea) with a 785 nm diode laser as the excitation source. The 1 mW of excitation light was focused on the sample through a 40 ×/0.6 NA objective with spot size ~2.4 μm. The spectrum of each point was measured 8 times in the range of 550 to 1500 cm^{-1} with a spectral resolution of 1 cm^{-1} and an integration time of 40 s at room temperature. The Raman spectrum was calibrated by measuring a silicon sample before the Raman measurements. To evaluate the spectral differences between control and IC/BPS of rat urine, principal component analysis (PCA) was introduced. A statistical analysis method of PCA reduces the number of variables in multivariate systems, and all of spectral range was used as variables. All analyses were conducted using XLSTAT 2018 software.

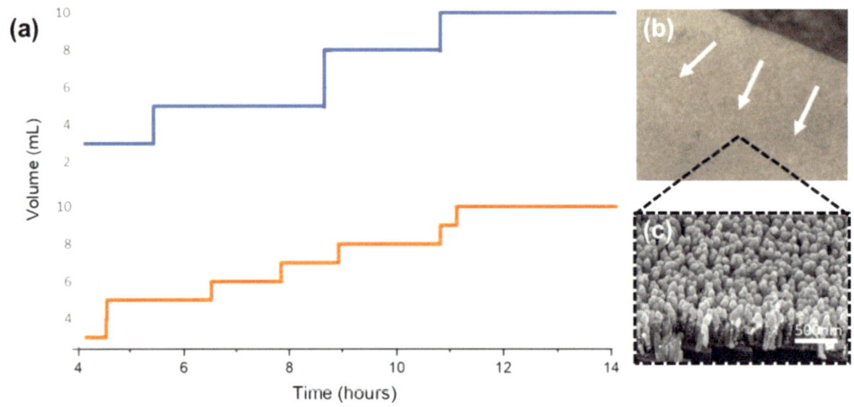

Figure 6.
(a) Measurement of voiding function in control group (blue line) and IC/BPS animal group (orange line) at 7 days after HCl treatment. (b) Optical microscope images of a Raman measurement region diffused from a sample droplet into a nanoporous area and (c) magnified SEM image.

As shown in **Figure 7a**, brown bars are indicated on the peak, which is the main factor above the graph drawn as the average of total data. The main peaks for the control and IP/BPS samples were observed at 641, 683, 723, 873, 1002, 1030, and 1355 cm^{-1}, which corresponded to C-C twisting mode of tyrosine [37, 38], ring breathing of nucleic acids for G [38, 39] and A [39, 40], C-C stretch of hydroxyproline [37, 41], symmetric ring breathing mode [37–41] and C-H in-plane bending mode of phenylalanine [37, 41], and CH_3CH_2 wagging mode of collagen [37, 41], respectively.

The peak at 1002 cm^{-1} has a considerably large value compared to the rest of the data, which is notable in literature referring to other peaks, as they relate to other biologies. To analyze Raman peaks, PCA is utilized as shown in **Figure 7b**. Clear discrimination and reliable separation between control and IC/BPS groups were observed. By plotting PC2 and PC3, the groups show clear distinctions between IC/BPS urine and the control samples (dotted line). Raman spectrum measurements and PCA analysis showed that it is possible to distinguish between normal and diseased groups using gold-coated ZnO nanorod substrates that can be applied to early disease diagnostic sensing chips.

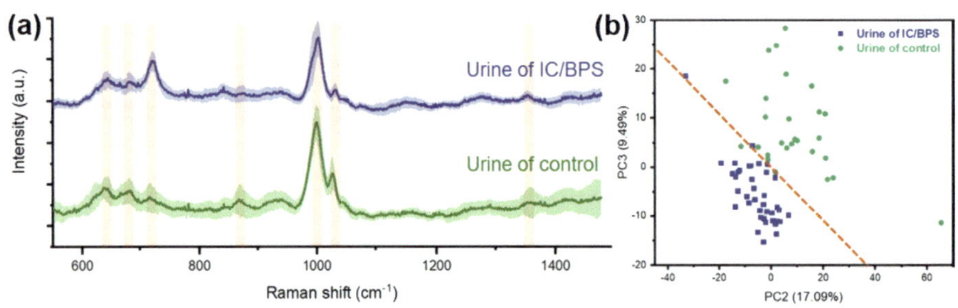

Figure 7.
(a) Averaged Raman spectra for IC/BPS (blue line) and control (green line) of rat's urine. Standard deviations are painted around the spectra. (b) Principal component analysis results for urine of IC/BPS and control sample.

7. Conclusion

In summary, we compared the differences in surface-enhanced Raman effect using RhB by adjusting the length and diameter of ZnO nanorods and the volume of deposited gold in ZnO-based SERS substrates. Electron microscopy images showed clustering on top of nanorods during gold deposition and showed nanometer level porosity. As the volume of deposited gold increases, the Raman signal also improves, but as the growth conditions of the nanorods change, the signal intensity also changes. This is because the Raman enhancement factor is determined by the enhancement by the SERS properties of metal and the concentration of the sample. Through the finite element analysis in the two-dimensional plane, signal enhancement was similar for 80 and 125 nm of Au-grain head. It was found that the enhancement of the Raman signal was determined by the wider surface area of gold. It was confirmed that this signal enhancement is made in the vertical direction of the rod, so that only nanometer targets trapped in the porous space can obtain enhanced signals. In addition, the SERS chips based on ZnO nanorods were found to have no coffee ring effect in measuring liquid samples and were also suitable for cell application experiments. An IC/BPS animal model was constructed for the bio-application, the voiding pattern was observed for urinary disease status, and urine was collected from the IC/BPS. The obtained urine was diffused into a ZnO-based SERS chip having nanopores, and Raman was measured in the corresponding region. Statistical analysis of Raman signals obtained from nanometric level area showed that IC/BPS and normal animals were distinguished. Therefore, we can confirm that ZnO nanorod-based SERS has sufficient potential for early disease diagnosis by efficiently detecting nano-sized biomarkers.

Acknowledgements

This work was supported by the Basic Science Research Program (2018R1D1A1B07048562) and MRC grant (2018R1A5A2020732) through the National Research Foundation of Korea (NRF) funded by the Ministry of Science and ICT (MSIT), by the Ministry of Trade, Industry and Energy (MOTIE) under the Industrial Technology Innovation Program (10080726, 20000843), and by a grant of the Korea Health Technology R&D Project through the Korea Health Industry Development Institute (KHIDI), funded by the Ministry of Health and Welfare, Republic of Korea (HI18C2391).

Conflict of interest

The authors declare no conflict of interest.

Author details

Sanghwa Lee[1] and Jun Ki Kim[1,2*]

1 Biomedical Engineering Research Center, Asan Institute of Life Sciences, Asan Medical Center, Seoul, Republic of Korea

2 Department of Convergence Medicine, University of Ulsan College of Medicine, Seoul, Republic of Korea

*Address all correspondence to: kim@amc.seoul.kr

References

[1] Ishigaki M, Maeda Y, Taketani A, Andriana BB, Ishihara R, Wongravee K, et al. Diagnosis of early-stage esophageal cancer by Raman spectroscopy and chemometric techniques. Analyst. 2016;**141**(3):1027-1033

[2] Kikuchi H, Kawabata T, Okazaki S, Hiramatsu Y, Baba M, Ohta M, et al. Near-infrared multichannel Raman spectroscopy with a 1064-nm excitation wavelength for ex vivo diagnosis of gastric cancer. Cancer Research. 2011;**71**

[3] Carvalho LFCS, Bonnier F, O'Callaghan K, O'Sullivan J, Flint S, Byrne HJ, et al. Raman micro-spectroscopy for rapid screening of oral squamous cell carcinoma. Experimental and Molecular Pathology. 2015;**98**(3):502-509

[4] Lee SH, Kim OK, Lee S, Kim JK. Local-dependency of morphological and optical properties between breast cancer cell lines. Spectrochimica Acta A. 2018;**205**:132-138

[5] Smith R, Wright KL, Ashton L. Raman spectroscopy: An evolving technique for live cell studies. Analyst. 2016;**141**(12):3590-3600

[6] Liu TY, Tsai KT, Wang HH, Chen Y, Chen YH, Chao YC, et al. Functionalized arrays of Raman-enhancing nanoparticles for capture and culture-free analysis of bacteria in human blood. Nature Communications. 2011;**2**(538):1-8

[7] Zhou HB, Yang DT, Ivleva NP, Mircescu NE, Niessner R, Haisch C. SERS detection of bacteria in water by in situ coating with Ag nanoparticles. Analytical Chemistry. 2014;**86**(3):1525-1533

[8] Khatun Z, Bhat A, Sharma S, Sharma A. Elucidating diversity of exosomes: Biophysical and molecular characterization methods. Nanomedicine-Uk. 2016;**11**(17):2359-2377

[9] Lee C, Carney RP, Hazari S, Smith ZJ, Knudson A, Robertson CS, et al. 3D plasmonic nanobowl platform for the study of exosomes in solution. Nanoscale. 2015;7(20):9290-9297

[10] Maiti NC, Apetri MM, Zagorski MG, Carey PR, Anderson VE. Raman spectroscopic characterization of secondary structure in natively unfolded proteins: Alpha-synuclein. Journal of the American Chemical Society. 2004;**126**(8):2399-2408

[11] Rygula A, Majzner K, Marzec KM, Kaczor A, Pilarczyk M, Baranska M. Raman spectroscopy of proteins: A review. Journal of Raman Specroscopy. 2013;**44**(8):1061-1076

[12] Stiles PL, Dieringer JA, Shah NC, Van Duyne RR. Surface-Enhanced Raman Spectroscopy. Annual Review of Analytical Chemistry. 2008;**1**:601-626

[13] Etchegoin PG, Le Ru EC. A perspective on single molecule SERS: Current status and future challenges. Physical Chemistry Chemical Physics. 2008;**10**(40):6079-6089

[14] Li WY, Camargo PHC, Lu XM, Xia YN. Dimers of silver nanospheres: Facile synthesis and their use as hot spots for surface-enhanced Raman scattering. Nano Letters. 2009;**9**(1):485-490

[15] Wang YL, Irudayaraj J. Surface-enhanced Raman spectroscopy at single-molecule scale and its implications in biology. Philos T R Soc B. 1611;**368**:2013

[16] Caldwell JD, Glembocki OJ, Bezares FJ, Kariniemi MI, Niinisto JT, Hatanpaa TT, et al. Large-area plasmonic hot-spot arrays: Sub-2 nm interparticle separations with plasma-enhanced

atomic layer deposition of Ag on periodic arrays of Si nanopillars. Optics Express. 2011;**19**(27):26056-26064

[17] Lee SJ, Morrill AR, Moskovits M. Hot spots in silver nanowire bundles for surface-enhanced Raman spectroscopy. Journal of the American Chemical Society. 2006;**128**(7):2200-2201

[18] Kleinman SL, Frontiera RR, Henry AI, Dieringer JA, Van Duyne RP. Creating, characterizing, and controlling chemistry with SERS hot spots. Physical Chemistry Chemical Physics. 2013;**15**(1):21-36

[19] Abu Hatab NA, Oran JM, Sepaniak MJ. Surface-enhanced Raman spectroscopy substrates created via electron beam lithography and nanotransfer printing. ACS Nano. 2008;**2**(2):377-385

[20] De Angelis F, Gentile F, Mecarini F, Das G, Moretti M, Candeloro P, et al. Breaking the diffusion limit with super-hydrophobic delivery of molecules to plasmonic nanofocusing SERS structures. Nature Photonics. 2011;**5**(11):683-688

[21] Cinel NA, Cakmakyapan S, Butun S, Ertas G, Ozbay E. E-Beam lithography designed substrates for surface enhanced Raman spectroscopy. Photonic Nanostruct. 2015;**15**:109-115

[22] Marotta NE, Barber JR, Dluhy PR, Bottomley LA. Patterned silver nanorod array substrates for surface-enhanced Raman scattering. Applied Spectroscopy. 2009;**63**(10):1101-1106

[23] Ngo YH, Li D, Simon GP, Gamier G. Paper surfaces functionalized by nanoparticles. Adv Colloid Interfac. 2011;**163**(1):23-38

[24] Li BW, Zhang W, Chen LX, Lin BC. A fast and low-cost spray method for prototyping and depositing surface-enhanced Raman scattering arrays on microfluidic paper based device. Electrophoresis. 2013;**34**(15):2162-2168

[25] Zhang W, Li BW, Chen LX, Wang YQ , Gao DX, Ma XH, et al. Brushing, a simple way to fabricate SERS active paper substrates. Anal Methods-Uk. 2014;**6**(7):2066-2071

[26] Schmidt MS, Hubner J, Boisen A. Large area fabrication of leaning silicon nanopillars for surface enhanced Raman spectroscopy. Advanced Materials. 2012;**24**(10):Op11-OOp8

[27] Albiss BA, AL-Akhras MA, Obaidat I. Ultraviolet photodetector based on ZnO nanorods grown on a flexible PDMS substrate. Int J Environ an Ch. 2015;**95**(4):339-348

[28] Chen Y, Tse WH, Chen LY, Zhang J. Ag nanoparticles-decorated ZnO nanorod array on a mechanical flexible substrate with enhanced optical and antimicrobial properties. Nanoscale Research Letters. 2015;**10**

[29] Ibupoto ZH, Khun K, Eriksson M, AlSalhi M, Atif M, Ansari A, et al. Hydrothermal growth of vertically aligned ZnO nanorods using a biocomposite seed layer of ZnO nanoparticles. Materials. 2013;**6**(8):3584-3597

[30] Sinha G, Depero LE, Alessandri I. Recyclable SERS substrates based on Au-coated ZnO nanorods. Acs Appl Mater Inter. 2011;**3**(7):2557-2563

[31] Tripp DA, Nickel JC, Wong J, Pontari M, Moldwin R, Mayer R, et al. Mapping of pain phenotypes in female patients with bladder pain syndrome/interstitial cystitis and controls. European Urology. 2012;**62**(6):1188-1194

[32] Evans RJ, Moldwin RM, Cossons N, Darekar A, Mills IW, Scholfield D. Proof

of concept trial of tanezumab for the treatment of symptoms associated with interstitial cystitis. J Urology. 2011;**185**(5):1716-1721

[33] Hanno PM, Erickson D, Moldwin R, Faraday MM. Diagnosis and treatment of interstitial cystitis/bladder pain syndrome: AUA guideline amendment. J Urology. 2015;**193**(5):1545-1553

[34] Nickel JC, Herschorn S, Whitmore KE, Forrest JB, Hu P, Friedman AJ, et al. Pentosan polysulfate sodium for treatment of interstitial cystitis/bladder pain syndrome: Insights from a randomized, double-blind, placebo controlled study. J Urology. 2015;**193**(3):857-862

[35] Kim A, Yu HY, Lim J, Ryu CM, Kim YH, Heo J, et al. Improved efficacy and in vivo cellular properties of human embryonic stem cell derivative in a preclinical model of bladder pain syndrome. Sci Rep-Uk. 2017;**7**

[36] Song M, Park J, Choo MS. Mesenchymal stem-cell therapy alleviates interstitial cystitis by activating Wnt signaling pathway. J Urology. 2015;**193**(4):E217-E21E

[37] Stone N, Stavroulaki P, Kendall C, Birchall M, Barr H. Raman spectroscopy for early detection of laryngeal malignancy: Preliminary results. Laryngoscope. 2000;**110**(10):1756-1763

[38] Synytsya A, Judexova M, Hoskovec D, Miskovicova M, Petruzelka L. Raman spectroscopy at different excitation wavelengths (1064, 785 and 532 nm) as a tool for diagnosis of colon cancer. Journal of Raman Specroscopy. 2014;**45**(10):903-911

[39] Chan JW, Taylor DS, Zwerdling T, Lane SM, Ihara K, Huser T. Micro-Raman spectroscopy detects individual neoplastic and normal hematopoietic cells. Biophysical Journal. 2006;**90**(2):648-656

[40] Wang H, Huang N, Zhao J, Lui H, Korbelik M, Zeng H. Depth-resolved in vivo micro-Raman spectroscopy of a murine skin tumor model reveals cancer-specific spectral biomarkers. Journal of Raman Specroscopy. 2011;**42**(2):160-166

[41] Huang ZW, McWilliams A, Lui H, McLean DI, Lam S, Zeng HS. Near-infrared Raman spectroscopy for optical diagnosis of lung cancer. International Journal of Cancer. 2003;**107**(6):1047-1052

Chapter 7

Anodic ZnO-Graphene Composite Materials in Lithium Batteries

Herrera-Pérez Gabriel, Pérez-Zúñiga Germán, Verde-Gómez Ysmael, Valenzuela-Muñiz Ana María and Vargas-Bernal Rafael

Abstract

An important area to cope with in the implementation of technologies for the generation of energy from renewable sources is storage, so it is a priority to develop new ways of storing energy with high efficiency and storage capacity. Experimental reports focused on ZnO-graphene composite materials applied to the anode design which indicated that they show low efficiencies of around 50 %, but values very close to the theoretical capacity have already been reported in recent years. The low efficiency of the materials for the anode design of the Li-ion battery is mainly attributed to the pulverization and fragmentation of the material or materials, caused by the volumetric changes and stability problems during the charge/discharge cycles. In this chapter, we will discuss the development of composite materials such as ZnO-graphene in its application for the design of the anode in the Li-ion battery.

Keywords: ZnO, batteries, graphene, Li-ion, composites, ZnO-Graphene

1. Introduction

The search for new and efficient energy storage systems has been mainly aimed at batteries, which have positioned themselves as one of the best options for this purpose [1]. The development and innovation in such systems has been slow compared to other technologies since the relevant innovations have taken even centuries, from the invention of the battery in 1800 by Alessandro Volta [2], then the first lead-acid batteries partially rechargeable in 1860 by Gaston Planté [3], up to the lithium-ion batteries, which Sony introduced to the market in 1991 [4].

The batteries can be classified as primary (not rechargeable) and secondary (rechargeable). The classification of the batteries is made in relation to the active materials, in the components of the battery. At present, there are still few batteries with the ability to have reversible electrochemical reactions. **Table 1** shows some of the main physical and chemical properties of current commercial batteries. The Li-ion batteries, clearly positioned today, as one of the best options for energy storage, are far above the other batteries in terms of the number of cycles.

1.1 Overview of batteries

The first Li-based batteries used $LiCoO_2$ cathode, anode carbon, and $LiPF_6$ electrolytes, with a capacity of around 140 mAh and 3.7 V and an efficiency of 50 %,

Battery type	Anode	Cathode	Electrolyte	Voltage (V)	Duration (cycles)
Li-ion	Graphite	$LiCoO_2$	$LiPF_{6(nonaqueous)}$	3.7	>1000
Pb-acid	Pb	PbO_2	$H_2SO_{4(aqueous)}$	2.1	<500
Ni-Cd	Cd	NiOOH	$KOH_{(aqueous)}$	1.2	2000
NMH	$Ti_2Ni+TiNi$	NiOOH	$KOH_{(aqueous)}$	1.2	500–1000

Table 1.
General characteristics of the most commonly used commercial batteries [5].

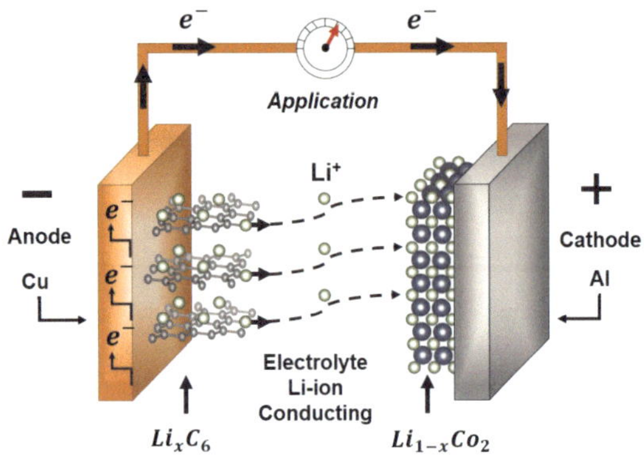

Figure 1.
Schematic of a Li-ion cell.

which was relatively low. The innovations that followed in the Li-ion batteries revolved around the three main components that are the anode, cathode, and electrolyte, in order to improve the characteristics and problems that currently accompany this technology, such as the capacity of gravimetric and volumetric power, safety, cost, efficiency, and the search for materials that do not harm the environment [4, 6, 7].

A Li-ion battery is a set of cells, connected in series or in parallel in **Figure 1**. A diagram of the main components of a Li cell is shown, which are a negative electrode or anode and the positive electrode or cathode and between them a Li-ion conductive electrolyte. The principle of the operation of a cell is based on the transport of Li-ions between the two electrodes, capable of storing Li [1].

During the charge and discharge processes, the electrons flow from the anode to an external load and to the cathode in a discharge process, through an oxidation reaction at the anode and a flow of Li-ions from the anode to the cathode through the electrolyte. In a charging process, the flow of electrons is from the cathode to a source and to the anode, followed by a reduction reaction at the cathode and a flow of Li-ion to the anode through the electrolyte, forming a circuit. The collectors of charge or substrates that are used regularly are copper and aluminum for the anode and cathode, respectively [8].

The standard potential of a cell is determined by the type of active materials in the cell and can be calculated in relation to the Gibbs free energy or obtained experimentally. This potential can be calculated from the standard potential of the electrodes in the following way:

$$\text{anode (oxi.potential)} + \text{cathode (red.potential)} = \text{standard cell potential} \quad (1)$$

The oxidation potential is the negative value of the reduction potential; it is worth mentioning that it is dependent on other factors such as concentration and temperature. The theoretical capacity of a cell is determined by the amount of active materials; this expresses the amount of electricity involved in the electrochemical reaction, which can be expressed in terms of coulombs or ampere/hour [8].

1.2 Materials for the design of electrodes

Within the characteristics of the electrodes, two phenomena mainly occur: diffusion and adsorption. The diffusion phenomenon happens in the cathode, and the carbon that is commonly used in the anode presents a phenomenon of adsorption. Nowadays the diversity of existing materials that are used or that are proposed for the electrodes and the great majority are governed by diffusion phenomena or in a composite material (CM) in which both phenomena are involved [9].

The first Li-ion batteries had been based on components as $LiCoO_2$ electrodes in the cathode and carbon in the anode. The development and innovation in the components of a rechargeable battery have considerably improved characteristics, such as the capacity of gravimetric and volumetric power, which has led to the development of batteries with equal or greater power but of lower weight and volume [6].

1.2.1 Cathode

A wide variety of materials has been used for the cathode since the introduction of the first Li-ion batteries. Specifically, Li-ion batteries are not governed by the chemical potentials of the materials involved, so the Li/Li^+ potential for both electrodes is taken as a reference. This means that the material of the cathode that will interact with the Li must have a high potential (+) in relation to that of Li/Li^+ [8, 9].

One of the most significant innovations that was made in the first batteries was the replacement of the Co, which was expensive and also toxic, being replaced by the Mn, which significantly decreased the price and at the same time increased the power capacity to 250 $mAhg^{-1}$ and a voltage of 4.6–2.5 V [6]. After the innovation that occurred with the $LiMn_2O_4$, new proposals like $LiFePO_4$ [10] were developed, and recently high ionic conductivity systems were investigated, in which the S is mainly involved, which has a theoretical capacity of 1675 $mAhg^{-1}$ [6].

1.2.2 Anode

The graphite that was used in the first Li-ion batteries had a capacity that was theoretically limited to 372 $mAhg^{-1}$ [4], and as a solution to these limitations, various materials have emerged such as Si, Sn, Sb, Ge, and new forms allotropic carbon [11]. The materials used at the anode, unlike the cathode, must have a potential no greater than that of Li/Li^+ and less than the potential of the cathode. That is, the materials that are used as electrodes should not have a greater potential than Li/Li^+ since the Li-ions could be reduced by forming metallic Li and in certain circumstances dendrites that considerably reduce the cycles and especially generate problems of safety since they can cause short circuit, which is one of the main problems when metallic Li is used as an anode [5, 8].

In recent years, transition metal oxides (TMO) have been used, such as Ni, Sn, Mn, and Zn, among others. In a fully charged Li battery, the anode contains an excess of Li-ions that have a chemical potential to diffuse through the electrolyte and into the vacancies of the cathode structure [5]. **Figure 2** shows a representative diagram of a charging and discharging process for a Li battery, and as can be seen

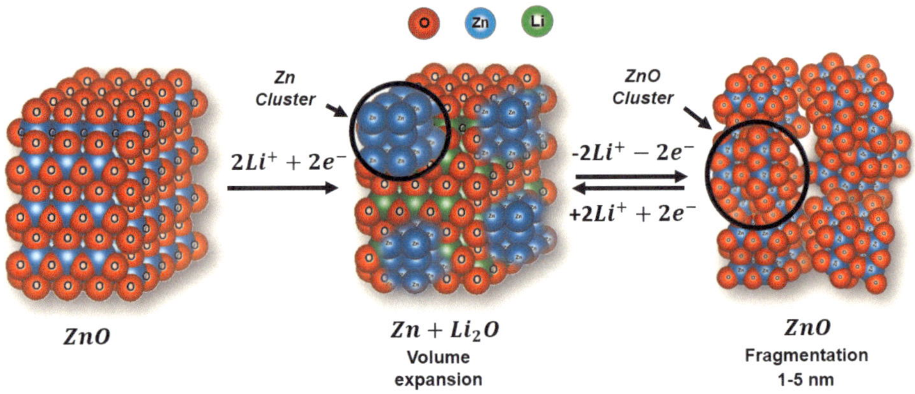

Figure 2.
Reversible conversion reaction of an metallic oxide (MO) with Li.

during the lithiation process, the metallic oxide (MO) is reduced to its metallic state, inside a Li_2O matrix according to the following reaction:

$$MO_x + 2xLi^+ + 2xe^- \leftrightarrow M + xLi_2O \qquad (2)$$

In the lithiation process, metal nanoparticles (M) embedded in a Li_2O matrix are formed (this generates volumetric expansion), and in the first discharge the nanoparticles of metal (M) are oxidized again into smaller particles.

Graphene is a material that in recent years has attracted a lot of attention due to its exceptional properties and its potential application in many areas, being one of them in Li-ion batteries, where it is used in the negative electrode. However, it has been seen that it cannot be used in a pure manner, because after the first discharge, it presents great problems of reversibility, so it is common for it to be widely used with other materials in a composite, generally with MO [5, 12].

2. Materials for the design of the ZnO-graphene system

Currently, graphene is one of the most researched and promising materials to replace graphite in Li-ion battery anode. When using graphene in an CM, it is expected to take advantage of the properties of high conductivity and high surface area, and in the specific case when used with an MO, it works as a buffer of the volumetric changes that it undergoes in lithiation [13]. One of the pioneers in the study of the intercalation of Li in various carbonaceous materials was Dahn et al. [14] where they concluded that graphite is limited to a certain amount of Li retention and that, in addition most of the processes of interaction with lithium, it is dominated by mechanisms of physical interaction.

Graphene, which is usually used as CM, usually comes from graphite and is obtained as GO. In several tests that have been carried out as anodes, irreversible capacities of up to 1250 $mAhg^{-1}$ have been obtained in the first cycle; this is due to the formation of a solid interface with the electrolyte (SEI), which is one of the main causes of the decrease in capacity, which at the same time is due to the high surface area [5].

Carbon/metal CMs are made with metals that are capable of forming alloys with Li or transition metal oxides (TMO), a term introduced by Tarascon et al. in the year 2000 and where they show several MO reporting reversible capacities of up to 700 $mAhg^{-1}$ [15]. One of the main disadvantages when using an MO is that they are generally poor electrical conductors or in certain cases semiconductors, being one of the causes for which they are commonly used in CM, with carbonaceous materials that compensate for these deficiencies of conductivity and volumetric changes [5].

One of the MO that is being investigated as an anode is ZnO, which theoretically has a capacity of 987 $mAhg^{-1}$; the charge and discharge reactions can be written as follows:

$$ZnO + 2Li^{+} + 2e^{-} \leftrightarrow Zn + Li_2O \tag{3}$$

$$Zn + Li^{+} + e^{-} \leftrightarrow LiZn \tag{4}$$

Complete reaction:

$$ZnO + 3Li^{+} + 3e^{-} \leftrightarrow LiZn + Li_2O \tag{5}$$

The theoretical capacity of ZnO can be calculated as follows:

$$\begin{aligned} C_t\left(\text{mAhg}^{-1}\right) &= \left(z \times F\left(\text{Cmol}^{-1}\right)/\left(M_w\left(\text{gmol}^{-1}\right)\right)\right) = \\ &= (\text{z} \times 96485\ (\text{A s}))/81.39\ (\text{g}) = \\ &= (\text{z} \times 26801\ (\text{mAh}))/(\text{M}_\text{w}(\text{g})) = \\ &= (3 \times 26801\ (\text{mAh}))/81.39\text{g}) = 987.87\ \left(\text{mAhg}^{-1}\right) \end{aligned} \tag{6}$$

where C_t is the theoretical specific capacity, F is the Faraday constant, z is the number of electrons transferred from each structural unit, and M_w is the molecular weight [16].

The theoretical capacity of ZnO is one of the reasons why this material is interesting to be used as an anode in Li-ion battery. However, it is not only of interest because of its theoretical capacity, since it is also a material that, unlike the other transition metal oxides, has certain advantages such as high chemical stability, low cost, nontoxic, and relative ease of synthesis with a large number of methods and a great variety of structures and morphologies [17–21].

2.1 Zinc oxide

ZnO is a material of great interest in various research areas and a large number of technological applications, such as in ceramics, piezoelectric, transducers, chemical sensors, catalysis, optical applications, photovoltaics, and lithium batteries, among many others. For this reason, the interest in this material continues to be latent since it is also one of the few oxides that shows effects of quantum confinement to certain particle sizes [22–24].

ZnO is a material with great versatility, in terms of synthesis methods and nanostructures, such as nanobars [20], nanosheets [25, 26], 3D nanostructures [27, 28], and nanocrystals [18, 29]. Within the synthesis methods, relatively complex methods are found such as the chemical vapor deposition (CVD) [30], Vapor Phase Transport (VPT) [31] for its acronym in English. These techniques are relatively expensive and complex, compared to others such as chemical precipitation [28, 32], sol-gel [23, 33–35], hydrothermal [29, 36–41], or solvothermal [19, 42, 43].

The Wurtzite phase that is defined by a hexagonal crystalline system is the most stable phase of ZnO at environmental conditions, but this ZnO can also be obtained in cubic phase forming the zinc blende; and using Si substrates and with high pressures, the rock of salt (NaCl) form is obtained. The structure of the orthorhombic unit cell of the main precursor phase of the Wurtzite, which is the Wülfingite, that chemically is $Zn(OH)_2$, is presented in **Figure 3a**, **b**. For wurzite form is presented in **Figure 3c**, **d** the Hexagonal crystalline system and that belongs to the space group P63mc, with the network parameters a = 0.35 nm and c = 0.52 nm [44]. Wurtzite is a material with a relatively simple atomic crystalline structure, in which it can be seen that each atom of Zn is surrounded by four atoms of O forming

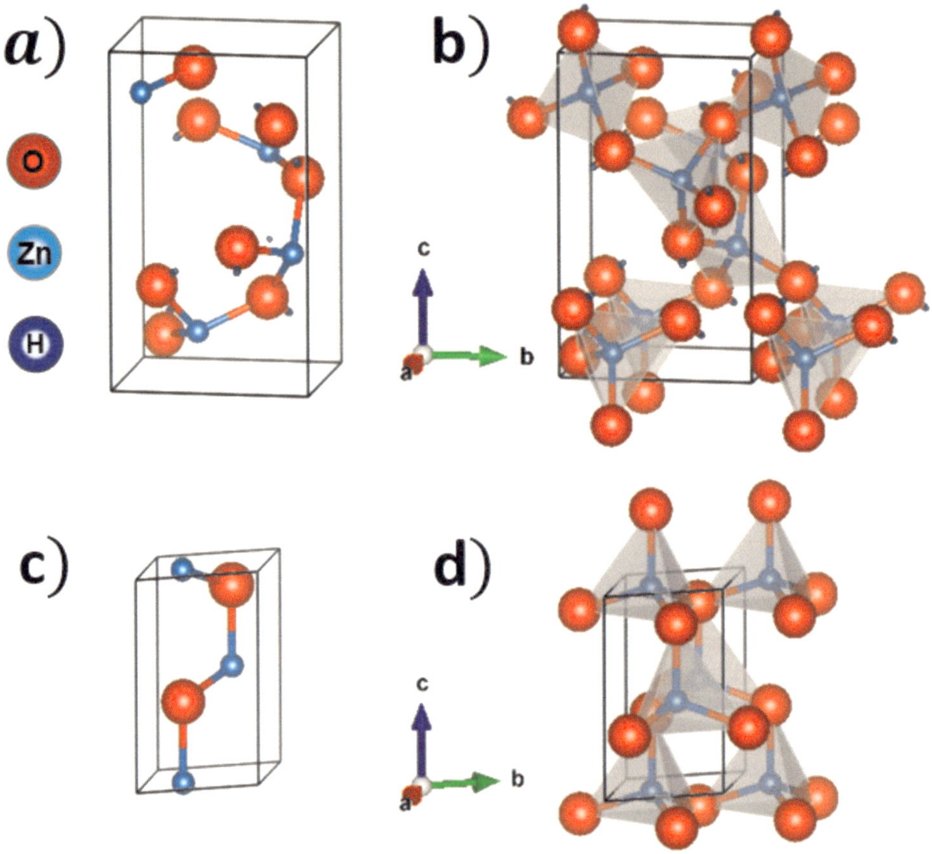

Figure 3.
(a) Unitary cell of ε-Zn(OH)$_2$, (b) representation of polyhedra of the unitary cell of the wülfingite, (c) unitary cell of ZnO, and (d) representation of polyhedra of the unit cell of the wurtzite.

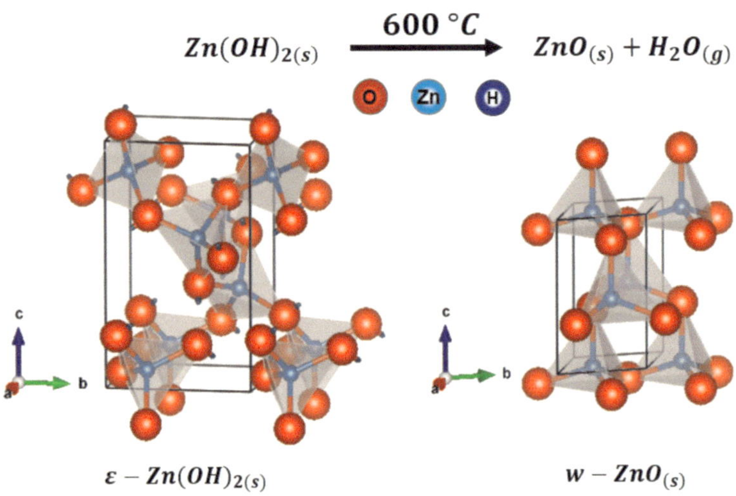

Figure 4.
Thermal transformation of zinc hydroxide (wülfingite) for zinc oxide (wurtzite).

a tetrahedron and likewise with each atom of O surrounded by four atoms of Zn forming tetrahedral indicating the one-to-one ratio Zn:O [24, 45].

Figure 4 shows a relatively simple transformation of a metal hydroxide to a metal oxide, by dehydroxylation of Wülfingite only Wurtzite can be obtained; in the scientific literature, there is a very extensive information on the relative simplicity of this transformation, this time a temperature of 600 °C is a maximum referent, but it is known that depending on the synthesis techniques this temperature can be reduced to promote a smaller crystal growth, as well as contribute to decrease the possible secondary processes at low temperatures this is between about 90 and 250 °C.

The interest in ZnO and its application as an anode in lithium-ion batteries is due to its great theoretical capacity (987 $mAhg^{-1}$) [25], this with respect to graphite that has a theoretical capacity of around 372 $mAhg^{-1}$, which is the most commonly used material as an anode in lithium batteries [46]. In addition, its theoretical capacity puts it ahead of other transition metals such as Cu, Ni, Fe, or Ti, and even on the Sn of group IV, where there are elements such as Ge and Si of greater theoretical capacities but that have certain disadvantages in terms of their relatively complex synthesis processes and in the case of Ge, rarity and toxicity [47].

Si is the material with the highest theoretical capacity (4200 $mAhg^{-1}$); however, one of the big problems is that in a few cycles (10 cycles), it loses its capacity until 200 $mAhg^{-1}$, due to the great volumetric expansion it undergoes in the lithiation process (~400 %) [47]. Similarly, ZnO undergoes volumetric changes of ~228 %, which is reflected in a decrease in its efficiency and capacity to ~400 $mAhg^{-1}$; however, the ZnO has the novelty of having a wide variety of synthesis processes, morphologies, and structures that could help solve these problems [20, 48] as shown in **Figure 4**.

The solution to these problems has motivated several research groups that have chosen to perform CM of ZnO with a great variety of carbonaceous materials, graphite [25], nanotubes [21], or graphene [32, 35, 49, 50]. When using carbonaceous materials, two problems with ZnO are compensated: the low electrical conductivity and as a buffer for the volumetric changes of ZnO [48, 50].

2.2 Graphene

Graphene is a crystalline, two-dimensional (2D) material of a single atom in thickness, with covalent bonds σ, sp^2 between carbon and carbon, with a junction

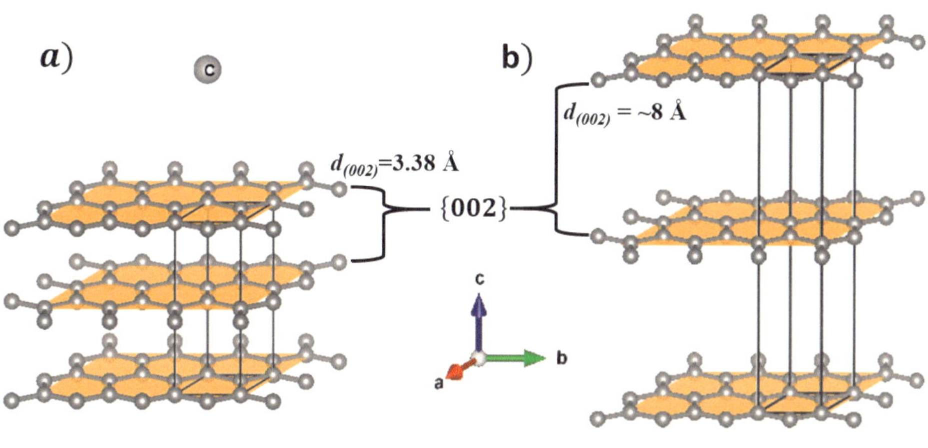

Figure 5.
Crystal structure of (a) graphite and (b) after the oxidation process, a structure equivalent to the GO.

angle of 120° and a length of $a = 0.14$ nm, resulting in a hexagonal honeycomb lattice, which can be simplified to a trigonal unit cell with two atoms per unit, as shown in **Figure 5**.

Graphene can have other allotropic forms at the same time, since it is the basic building structure of the other forms such as graphite (3D), nanotubes (1D), and fullerenes (0D). In all its structures, it has very good properties of thermal, electrical, and other properties such as mechanical, optical, and magnetic properties. The movement of electrons through a sheet of graphene is governed mainly by the relativistic law of Dirac. For a perfect sheet of graphene, it is estimated that its conductivity reaches 200,000 cm^2/Vs; however, this intrinsic property is greatly affected by defects that generate dispersion centers generated mainly by substrates, dopants, and quiralities.

Since the nineteenth century, oxidized graphite has been produced by different methods: the Brodie method in 1859 [51], the Hummers method in 1958 [52], and the Hummers method modified in July 2004; and in October of the same year, Andre Geim and Kostya Novoselov, professors at the University of Manchester, London, managed to isolate a sheet of graphite a few atoms thick, by mechanical exfoliation, from which they discovered the exceptional properties of this material that has attracted enormous attention since then [53]. However, the synthesis of graphene remains a great challenge, since, to be considered high purity, it is necessary to take into account certain characteristics such as the quality, size, quantity, complexity, and control of the synthesis method. For this reason, it is very important to know the different methods, degrees of purity, and qualities expected for graphene [54]. In the work presented by Hirata et al. [55], they obtained sheets of several nanometers thick and 20 μm wide on average. In the same work, they propose several categories of graphene: oxidized graphene, reduced graphene, functionalized graphene, and graphene in its pure state.

3. Methods of synthesis of ZnO

The most used synthesis processes for ZnO are chemical precipitation at normal conditions, synthesis by hydrothermal, solvothermal techniques, and the sol-gel method. The advantage and novelty of these techniques is the obtaining of a great diversity of geometries, with relatively simple processes and with very accessible precursors.

3.1 Chemical precipitation at normal conditions

The technique is used specifically for ZnO, part of a Zn precursor, normally a salt or inorganic compound, soluble in water commonly and as a reducing agent either an acid or a base, which, when reacting in solution with the salt, will form a precipitate or solid of insoluble that will commonly require a calcination process for crystallization. This type of reactions could be expressed in a general equation in the following way:

$$AC + BD \rightarrow AD_{(\text{insoluble})} + BC \tag{7}$$

This type of reactions usually occurs between ionic compounds where one of the products is insoluble, and because each component exchanges pairs, these types of reactions are usually called double-displacement reactions.

Several authors such as Kundu et al. [20] and Liu et al. [48] obtained from chemical precipitation at normal conditions nanobars and ZnO nanoparticles, respectively. According to these same authors, they obtained the nanoparticles or nanobars individually or separately, and in the case of Lui et al. [48], they performed an extra process of carbon coating or as Giri et al. [28] who performed a hydrothermal procedure to coat the nanoparticles. On the other hand, Ramadoss et al. [32] obtained an CM where they precipitated nanoparticles of ZnO in situ, on graphene sheets.

3.2 Hydrothermal method

The method or technique of hydrothermal synthesis is one of the most used for the synthesis of ZnO, and with the same principle of the technique of chemical precipitation, this technique gets its name because the conditions of synthesis, which is carried out in water and temperatures above 25 °C up to $\sim$ 430 °C, reaching autogenous pressures of over 221 bar at $\sim$375 °C, which is the critical point where the liquid phase and gas are not distinguished from water [56]. This type of synthesis is also governed by the precipitation equation but adding temperature and pressure variables.

In various works such as that of Alver et al. [40], they synthesized ZnO doped with boron by the hydrothermal method and subsequently formed a nanocomposite with graphene using the same technique. Yoo et al. [41] with the same method obtained hemispherical nanoparticles of $\sim$25 nm. Wang et al. [57] synthesized ZnO in flower form, doped with Mn, with the hydrothermal method. Bøjesen et al. [36] conducted an in situ study of the growth of ZnO nanoparticles by hydrothermal synthesis. In this work, they conclude that temperature is one of the biggest factors that influences the size and shape characteristics of glass, this, without forgetting other factors such as time.

3.3 Solvothermal method

The solvothermal synthesis method is a variant of the hydrothermal one, since, in this technique, solvents different to water are used, but with the same principles of hydrothermal synthesis and governed by the precipitation equation. The said solvents can be alcohols, acids, bases, or mixtures, since, with this, a greater dissolution or changes in the pressures generated in the synthesis are sought [56]. With the solvothermal synthesis technique, authors such as Wi et al. [19] and Wang et al. [42] have obtained nanoparticle agglomerates, of spherical shape and porous sheet type, respectively. In the case of Wang et al. [42], the synthesis process was carried out at room temperature reaching thicknesses of up to 10 nm in the sheets.

3.4 Sol-gel

On the other hand, the sol-gel synthesis technique consists of a chemical synthesis in which, from a colloidal solution or "sol", small precipitates of a solid phase gradually form inside a continuous network called "gel". The peculiarity of the technique is the nanometric size of the particles that can be obtained by this technique as shown in the work of Spanhel et al. [32], who obtained ZnO nanoparticles of colloidal sizes between 3 and 6 nm. Hjiri et al. [33] obtained sizes between 20 and 40 nm for ZnO and up to 3 nm for ZnO doped with Al. Li et al. [34] used the technique to obtain in situ a nanocomposite of nanoparticles deposited in sheets of graphene, whose reported particle size was an average of $\sim$9 nm.

4. Graphene synthesis methods

To date, various synthesis methods have been developed for graphene, but basically, there are only two ways: one of them is to obtain the sheets of an existing crystal of graphite which is known as exfoliation methods, and on the other hand, the sheets of graphene can be grown directly on a substrate.

4.1 Exfoliation

The exfoliation methods can be classified in two: the micromechanical exfoliation method and the chemical methods, the latter can be by dispersion or oxidation. According to what was reported by Novoselov et al. [53], they used the technique of micromechanical exfoliation, obtaining sheets of graphene up to a single layer; however, this technique is relatively complex, and the yield or number of sheets is very low.

The dispersion method allows obtaining a greater amount of graphene and higher quality than the micromechanical exfoliation. The synthesis of graphene by oxidation is similar to that of dispersion since it is also in the liquid phase, but the graphene that is obtained is of a slightly lower quality, and this is because the GO has a large amount of defects and oxygen at the edges compared to the pristine [54, 57]. To improve the quality of the GO, it can undergo reduction processes and improve the quality of the graphene sheets, obtaining what was called reduced graphene sheets (rG) [54].

The method of Hummers and Offeman [52], reported since 1958, is a method to obtain large quantities of what they called oxidized graphite; in 2004, Hirata et al. [55] modified this method obtaining thinner sheets of better quality. To be reduced, either by heat treatments or chemical means, you can decrease the amounts of oxygen or eliminate it completely under certain conditions. By eliminating oxygen, the graphene sheets increase considerably their electrical conductivity, since the sp^3 bonds decrease while the sp^2 bonds increase, which attribute the excellent properties to graphene. The GO has defects associated with oxygen bonds, at the edges as well as within the plane, forming different functional groups as carbonyls (C = O), hydroxyls (—OH), and epoxy groups (—O—); in **Figure 6** a representative scheme of the structure of a GO sheet is shown; this model was proposed by Santos et al. [59] In addition, this method allowed the production of graphene at higher

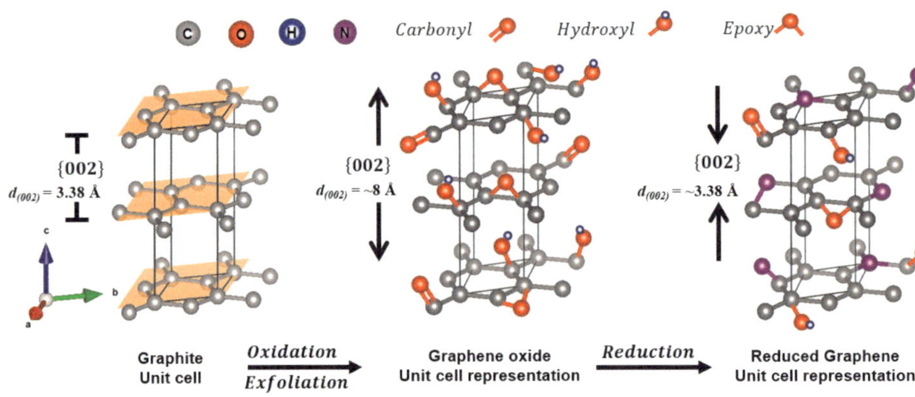

Figure 6.
Insertion or formation of functional groups in the sheets of graphene.

quantities and the possibility of being used industrially in different carbon products, one of which being Li-ion batteries [58, 59].

The modified Hummers method is currently one of the most used to obtain graphene, due to the sufficient quantity and the relative ease of the process, that once the GO is reduced, its properties are very favored, so that different groups of work have pointed their investigations using this technique to obtain graphene with applications for the anode of Li-ion batteries [61–63].

The Hummers method modified for the synthesis of graphene consists in exposing the graphite particles to prolonged times of combined oxidation to washing and purification processes. The process of Hirata et al. [55] is described as follows: 10 g of natural graphite (99.97 % purity and 24 μm particle diameter) and 7.5 g of $NaNO_3$ (99 % purity) were added to a flask. Then 621 g of H_2SO_4 (96 %) were added and kept stirring while it was cooled in an ice water bath. Next 45 g of $KMnO_4$ (99 %, purity) were added and gradually added over 1 h. The cooling was reached after 2 h, and the mixture remained for 5 days under constant agitation at 20 °C to obtain a highly viscous liquid.

The cleaning and purification process was carried out in the following way; to the obtained liquid, 1 L of solution with 5 % of H_2SO_4 was added during 1 h, while it was maintained in agitation, and the obtained mixture in the end was agitated during 2 h more. Then 30 g of H_2O_2 (30 % by weight solution) were added and kept under stirring for 2 h and finally separated by centrifugation. This process was repeated 15 times, for the removal of the ions originated in the oxidation process. The synthesis process used by Vargas et al. [64] has modifications such as the change in the use of $NaNO_3$ by HNO_3 and some modifications in synthesis and cleaning methodology. The process they used was the following: 2 g of graphite powder was added to an aqueous solution with 120 mL of H_2SO_4 and 80 mL of HNO_3 in an ice water bath. Then, 10 g of $KMnO_4$ was added slowly and remained in reaction for 2 h, during which the temperature remained around 35 °C. The dark brown suspension obtained was diluted with 400 mL of deionized water and turned dark yellow, to then add 8.6 mL of H_2O_2 (35 %). A dark brown gel was obtained, after washing with 100 mL of a 10 % solution of HCl and neutral pH obtained after several cycles of washing and centrifugation. The GO was finally obtained by vacuum drying at 80 °C.

Wan et al. [65] synthesized graphene using the modified Hummers method by applying the following process: 1 g of natural graphite was placed in a mixture with a concentrated solution of H_2SO_4 (98 %, 92 mL) and concentrated HNO_3 (65 %, 24 mL) while stirring in an ice water bath, as a safety measure and maintaining the temperature below 10 °C. Then 6 g of $KMnO_4$ was added to the solution gradually so that the temperature of the solution did not exceed 20 °C; then the solution was left for 2 h in the ice water bath. Then a temperature controller was used for water flow, to maintain the temperature at 35 °C for 30 min. Subsequently, the temperature was maintained at 85 °C for 30 min. About 100 mL of deionized water was slowly added to the solution, and the temperature was again maintained at 85 °C for 30 min, until a bright yellow product was obtained. After cooling to room temperature and diluting with 10 mL of 30 % H_2O_2. The solution was centrifuged, washed with 1 L of deionized water and a 1:10 HCl solution (1 L) to remove the remaining metal ions, and finally dried at 50 °C and vacuum for 12 h.

Of the three processes described above, the modified Hummers method differs in several aspects of the process such as the change of $NaNO_3$ by HNO_3 and the increase in temperature to reduce the oxidation time, as well as the optimization of the process of cleaning and obtaining the GO. According to the results obtained in what was reported by Wan et al. [65], they obtained GO of five sheets of thickness, with very promising results in the electrochemical tests they carried out.

4.2 Substrate growth

This is a totally different way to the previous methods since the graphene sheets can be grown directly on a surface. And the size of the sheets does not depend directly on the size of a graphite crystal, as in the exfoliation methods. Growth can occur in two ways: whether the carbon exists on the surface of the substrate or it is added by chemical vapor deposition (CVD). Graphene can be obtained simply by heating and cooling a silicon carbide crystal, under suitable conditions, obtaining sheets up to a single layer [54].

The chemical vapor deposition method is perhaps one of the most promising and relatively low-cost techniques to obtain good quality graphene. Broadly speaking, the technique consists in the deposition of a solid film on a substrate, where the chemical species of the material deposited come from species in vapor phase and are deposited through chemical reactions. In an ideal CVD process, the transport kinetics of gases is often complicated and complex, since convection and diffusion phenomena occur in different regions of a reactor [66].

The process for obtaining graphene by CVD can be divided into two stages: the first is the pyrolysis of the carbon precursor and the second the formation of the graphene structure. In an ideal synthesis to obtain graphene, temperatures of up to 2500 °C would be needed to overcome the energy barrier that allows the reaction on the surface of the substrate; for this reason, catalysts are used, which are mostly elemental metals, which contribute to the pyrolization of carbon precursors. One of the most used substrates is Ni (111) since it has a structure very similar to that of graphene, with a mismatch in network parameters of 1.3 % [54, 64].

The CVD process, as previously described, is a relatively complex process because of the equipment necessary to carry out the synthesis, but it allows obtaining graphene of higher quality at low cost, with larger sizes and more complex forms than the exfoliation processes of graphite [54, 57, 65, 66]. The team of Kim et al. [67] was among the first to report obtaining graphene by the CVD method on a Ni substrate, proving that the monolayers obtained were of much better quality than those of exfoliation. Since then, several authors have continued research to improve the technique, either by lowering the synthesis temperature as reported by Jang et al. [68], who obtained graphene at temperatures between 100 and 300 °C deposited on copper sheets, using benzene. Other authors such as Sagar et al. [58] have synthesized highly porous structures based on interconnected sheets of graphene and have proposed an anode in Li-ion batteries.

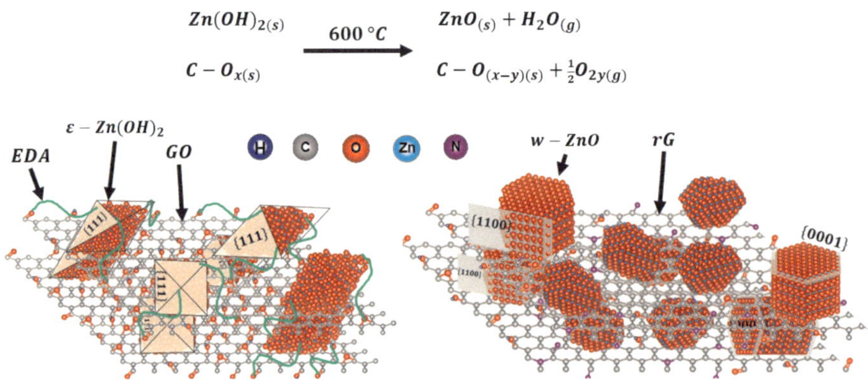

Figure 7.
Scheme of obtaining the MC type ZnO-rG. Left: material of the precursor phase $Zn(OH)_2$-GO. Right: the MC ZnO-rG obtained after a thermal treatment at 600 °C.

4.3 Composite ZnO-graphene materials

An CM of ZnO-graphene (ZnO-G) could have a large number of assembly structures, however, can be considered six principally, the most used in this type of CM for Li-ion batteries: (a) anchored, (b) wrapped, (c) encapsulated, (d) type sandwich, (e) laminar, and (f) mixed. In all cases, graphene, being a two-dimensional material, is the one that functions as a support for dispersed nanoparticles whose three-dimensional morphology can vary in different sizes, shapes, and crystallinity [13, 66].

Currently, the CM based on ZnO and carbon are very diverse and with variations in the morphology of both phases of the components. **Figure** 7 shows the obtaining of an CM of possible interest with the specific case of obtaining the dispersed phase of the phases of the ZnO with a Wurtzite type crystal structure, whose structure is more stable to standard conditions; this phase is the most attractive for, it is mainly used in CM for Li-ion batteries, also presented with a wide variety of morphologies; while for graphene, it usually presents different characteristics and properties according to its synthesis method [25, 69].

In recent years, a great amount of research has been carried out regarding the morphology of ZnO, since most of these works seek to increase the area, modifying the morphology and thus increasing the electrochemical properties, for its application in Li-ion batteries [16]. The investigations that have been carried out regarding the control of the crystallinity in the particles of the ZnO, to increase the capacity, are very few. Recently Mei et al. [17] published a paper in which they analyzed the degree of crystallinity and structural patterns of ZnO as an anode in Li-ion batteries. In this work, they used the hydrothermal synthesis in which they modified concentrations, to obtain different morphologies and then carried out treatments at different temperatures, this to control the degree of crystallinity. Finally, Mei et al. concluded that morphology was an important part in the capacity, since certain morphologies present a greater quantity of internal spaces, which help to compensate the volumetric changes; however, the samples whose crystallinity was controlled presented a specific capacity of 1328 $mAhg^{-1}$ in the first cycle and 663 $mAhg^{-1}$ at 50 cycles. In this work, it is worth mentioning that the particles obtained are of the micrometric order, contrary to what is generally reported with nanometric materials, such as Li et al. [18] that dispersed ZnO nanoparticles in

Morphology	Reversible capacity ($mAhg^{-1}$)	Number of cycles	Ref.
Microparticles distributed radially	320	100	[6]
Nanobars	310	40	[20]
Microbars	663	50	[17]
Nanoplates	400	100	[25]
Nanoplates	421	100	[25]
Nanoparticles	318	100	[42]
Flower type	662	50	[17]
Spheres	109	100	[70]
Cluster	800	50	[71]
Microbars	811	80	[72]
Nanoparticles	318	100	[73]

Table 2.
Electrochemical characteristics of ZnO applied as an anode material in Li-ion batteries.

graphene, obtaining an initial capacity of 1652 $mAhg^{-1}$, but the retention capacity decreased to 318 $mAhg^{-1}$.

Table 2 shows the data reported in the literature for the comparison of the reversible capacity and the number of cycles of the different morphologies of ZnO as an anode in Li-ion batteries. It is possible to identify the effect of each morphology in the particle of the ZnO; the geometry of the bars is distinguished from the others, because it presents a greater reversible capacity, which suggests a greater stability in the processes of lithiation and sliding. In **Table 2**, it can be seen that some reversible capacity values greater or closer to 600 $mAhg^{-1}$, the ZnO particles are CM shaped with a carbonaceous material. However, the value of 663 $mAhg^{-1}$ of the microbars reported by Mei et al. It is of great interest, because in spite of not being of nanometric size and not being like CM it has a great reversible capacity.

The ZnO-rG CMs promise better results due to the good properties of graphene. **Table 2** shows results of some CMs of ZnO with carbon, porous carbon, graphite,

Method	Initial capacity ($mAhg^{-1}$)	Reversible capacity ($mAhg^{-1}$)	Number of cycles	Ref.
Rapid exfoliation at 1050 °C/N_2	2000	1200	5	[64]
Exfoliation 300 °C/Ar	2100	600	100	[64]
Reduction with autoclaved N_2H_4 (DODA-Br)	3000	1100	50	[64]
Fast exfoliation at 1050 °C/Ar	2035	848	40	[5]
Exfoliation at 300 °C/Ar	2137	478	100	[65]
Thermal reduction exfoliation	1480	500	60	[74]
Thermal reduction exfoliation/N_2	3250	1354	50	[75]

Table 3.
Electrochemical characteristics of graphene in Li-ion batteries.

CM	Reversible capacity ($mAhg^{-1}$)	Number of cycles	Ref.
ZnO/G	516	100	[35]
ZnO/GN	1100	100	[50]
AZO/G	391	100	[49]
ZnO/G	360	200	[76]
ZnO/G	550	100	[77]
ZnO/G	560	100	[78]
ZnO/G	300	50	[79]
ZnO/CN	1047	100	[80]
ZnO/G	749	100	[81]
ZnO/G	550	100	[82]
Cu/ZnO/G	630	100	[83]
ZnO/C	520	100	[84]
ZnO/CN	1177	100	[85]
ZnO/NSG	720	100	[73]

NC = nitrided carbon; G = graphene; NG = nitrided graphene; NSG = nitrided-sulfurized graphene.

Table 4.
Electrochemical characteristics of an CM of ZnO-G in Li-ion batteries.

and graphene, of which graphene, in this case, is reduced with nitrogen and showed a very remarkable reversible capacity, comparing it with other results. The graphene used in most CMs is obtained by chemical exfoliation, called GO, which has a number of advantages with pristine-type graphene. The GO can be obtained in large quantities and of sufficient quality, using the modified Hummers method, to be used in CM in a large number of applications, as in Li-ion batteries [5, 54].

The high area and the conductivity properties of graphene are the main properties that call attention to this material to be used as an anode in Li-ion batteries. The first reported evidence of Li storage on a large scale was 500 $mAhg^{-1}$ for the first cycle and 300 $mAhg^{-1}$ at just 20 cycles; currently values of up to 3000 $mAhg^{-1}$ have been reported for the first cycle and decrements to values of 200 $mAhg^{-1}$ after a few cycles [64]. **Table 3** shows some data of the reported GO properties by different reduction methods. It should be mentioned that the GO of the data in **Table 3** was obtained by the modified Hummers method.

When analyzing the data in **Table 4**, it is observed that the efficiency or loss of capacity in graphene is notoriously greater due to the formation of the SEI; furthermore, it is observed that the GO reduced by thermal treatments, under an argon atmosphere, it has the best reversible capabilities and a greater number of cycles. The recently published ZnO-rG CMs are based on a reduced GO afterward and in which the ZnO nanoparticles are deposited or grow in situ in the leaves. In **Table 4**, some relevant values of CM of ZnO-rG are shown, in which the ZnO particles are of the order of hemispherical nanoparticles. Now, by comparing the values of the reversible capacity, it can be seen that they are close to 500 $mAhg^{-1}$ on an average of per 100 cycles. The data reported in the literature for ZnO-G composites have very low efficiencies of around 50 %, however, some reports such as that of Dai et al. [50], Eliana et al. [25], and Shen et al. [43] have reported values very close to the theoretical capacity.

The efficiency problems that currently arise in the materials for Li-ion battery anodes are mainly due to the pulverization and fragmentation of the said materials, caused by the volumetric changes and stability problems during the charge/discharge cycles. On the other hand, one of the factors that remain unfinished for this type of CMs is the surface interaction between graphene and MO that during the charge and discharge cycles is affected by parasitic reactions that impede the stability and diffusion of charges between both materials [13, 66].

5. Conclusions

The evidences reported in the scientific literature show an open thematic for the design of composite materials with the intention of directly impacting on this time of tool for the storage of electrochemical energy, being the ZnO-graphene system a great candidate to be used as a constituent material of the anode.

Acknowledgements

The authors are grateful for the support granted in the Call 2019 issued by the Program of Support for Scientific and Technological Research in the Educational Programs of the Federal, Decentralized Technological Institutes and Centers belonging to the National Technological of Mexico (TecNM) for the support assigned to the Project (5269.19-P); and to the Higher Technological Institute of Irapuato (ITESI) for the allocation of support from its Institutional Program for the Strengthening of Academic Staff (PIFOCA/PIFOPA/PIICYT 2019).

Author details

Herrera-Pérez Gabriel[1*], Pérez-Zúñiga Germán[2], Verde-Gómez Ysmael[2], Valenzuela-Muñiz Ana María[2] and Vargas-Bernal Rafael[1]

1 Department of Materials Engineering, Higher Technological Institute of Irapuato, Guanajuato, Mexico

2 Postgraduate Studies and Research Division, Technological Institute of Cancún, Quintana Roo, Mexico

*Address all correspondence to: gaherrera@itesi.edu.mx

References

[1] Scrosati B, Garche J. Lithium batteries: Status, prospects and future. Journal of Power Sources. 2010;**195**: 2419-2430

[2] Trasatti S. 1799–1999: Alessandro Volta's 'electric pile. Journal of Electroanalytical Chemistry. 1999;**460** (1–2):1-4

[3] Kurzweil P. Gaston Planté and his invention of the lead-acid battery-the genesis of the first practical rechargeable battery. Journal of Power Sources. 2010;**195**(14):4424-4434

[4] Armand M, Tarascon J-M. Building better batteries. Nature. 2008; **451**(7179):652-657

[5] Lockwood JD. Nanotechnology for Lithium-Ion Batteries. Boston, MA: Springer, US; 2013

[6] Amine K, Kanno R, Tzeng Y. Rechargeable lithium batteries and beyond: Progress, challenges, and future directions. MRS Bulletin. 2014;**39**(5): 395-401

[7] Robinson AL, Janek J. Solid-state batteries enter EV fray. MRS Bulletin. 2014;**39**(12):1046-1047

[8] Linden D, Reddy T. Handbook of Batteries, 3rd ed.; 2001

[9] Park M, Zhang X, Chung M, Less GB, Sastry AM. A review of conduction phenomena in Li-ion batteries. Journal of Power Sources. 2010;**195**(24): 7904-7929

[10] Gon J, Son B, Mukherjee S, Schuppert N, Bates A, Kwon O, et al. A review of lithium and non-lithium based solid state batteries. Journal of Power Sources. 2015;**282**:299-322

[11] Armstrong MJ, O'Dwyer C, Macklin WJ, Holmes JD. Evaluating the performance of nanostructured materials as lithium-ion battery electrodes. Nano Research. 2014;**7**(1): 1-62

[12] Fan X, Zheng WT, Kuo JL, Singh DJ. Adsorption of single li and the formation of small Li clusters on graphene for the anode of lithium-ion batteries. ACS Applied Materials & Interfaces. 2013;**5**(16):7793-7797

[13] Mazar Atabaki M, Kovacevic R. Graphene composites as anode materials in lithium-ion batteries. Electronic Materials Letters. 2013;**9**(2):133-153

[14] Dahn JR, Zheng T, Liu Y, Xue JS. Mechanisms for Lithium insertion in carbonaceous materials. Science. 1995; **270**(5236):590-593

[15] Tarascon J, Poizot P, Laruelle S, Grugeon S, Dupont L. Nano-sized transition-metal oxides as negative-electrode materials for lithium-ion batteries. Nature. 2000;**407**(6803): 496-499

[16] Yuan GH, Wang G, Wang H, Bai JT. Synthesis and electrochemical investigation of radial ZnO microparticles as anode materials for lithium-ion batteries. Ionics (Kiel). 2015;**21**(2):365-371

[17] Xiao L, Mei D, Cao M, Qu D, Deng B. Effects of structural patterns and degree of crystallinity on the performance of nanostructured ZnO as anode material for lithium-ion batteries. Journal of Alloys and Compounds. 2015; **627**:455-462

[18] Li H, Wei Y, Zhang Y, Yin F, Zhang C, Wang G, et al. Synthesis and electrochemical investigation of highly dispersed ZnO nanoparticles as anode material for lithium-ion batteries. Ionics (Kiel). 2016;**22**(8):1387-1393

[19] Wi S, Woo H, Lee S, Kang J, Kim J, An S, et al. Reduced graphene oxide/ carbon double-coated 3-D porous ZnO aggregates as high-performance Li-ion anode materials. Nanoscale Research Letters. 2015;**10**:1-8

[20] Kundu S, Sain S, Yoshio M, Kar T, Gunawardhana N, Pradhan SK. Structural interpretation of chemically synthesized ZnO nanorod and its application in lithium ion battery. Applied Surface Science. 2015;**329**: 206-211

[21] Guler MO, Cetinkaya T, Tocoglu U, Akbulut H. Electrochemical performance of MWCNT reinforced ZnO anodes for Li-ion batteries. Microelectronic Engineering. 2014;**118**: 54-60

[22] Wróbel J, Piechota J. On the structural stability of ZnO phases. Solid State Communications. 2008;**146**(7–8): 324-329

[23] Meulenkamp EA. Synthesis and growth of ZnO nanoparticles. The Journal of Physical Chemistry. B. 1998; **5647**(98):5566-5572

[24] Wang ZL. Zinc oxide nanostructures: Growth, properties and applications. Journal of Physics. Condensed Matter. 2004;**16**: R829-R858

[25] Quartarone E, Asta VD, Resmini A, Tealdi C, Tredici IG, Tamburini UA, et al. Graphite-coated ZnO nanosheets as high-capacity, highly stable, and binder-free anodes for lithium-ion batteries. Journal of Power Sources. 2016;**320**:314-321

[26] Zhang L, Zhao J, Lu H, Li L, Zheng J, Li H, et al. Facile synthesis and ultrahigh ethanol response of hierarchically porous ZnO nanosheets. Sensors and Actuators B: Chemical. 2012;**161**(1):209-215

[27] Yue H, Shi Z, Wang Q, Cao Z, Dong H, Qiao Y, et al. MOF-derived cobalt-doped ZnO@C composites as a high-performance anode material for lithium-ion batteries. ACS Applied Materials & Interfaces. 2014;**6**(19): 17067-17074

[28] Giri A, Pal P, Ananthakumar R, Jayachandran M, Mahanty S, Panda A. 3D hierarchically assembled porous wrinkled paper-like structure of $ZnCo_2O_4$ and Co-ZnO@ C as anode materials for lithium-ion batteries. Crystal Growth & Design. 2014;**14**: 3352-3359

[29] Søndergaard M, Bøjesen ED, Christensen M, Iversen BB. Supporting information size and morphology dependence of ZnO nanoparticles synthesized by a fast continuous flow hydrothermal method. Crystal Growth & Design. 2011;**11**(9):4027-4033

[30] Sun Y, Yang GZ, Cui H, Wang J, Wang CX. $Zn_xGe_{1-x}O$ 3D micronano structures with excellent performance as anode material in Lithium ion battery. ACS Applied Materials & Interfaces. 2015;7(28): 15230-15239

[31] Heo YW et al. ZnO nanowire growth and devices. Materials Science and Engineering R: Reports. 2004; **47**(1–2):1-47

[32] Ramadoss A, Kim SJ. Facile preparation and electrochemical characterization of graphene/ZnO nanocomposite for supercapacitor applications. Materials Chemistry and Physics. 2013;**140**(1):405-411

[33] Spanhel L, Anderson MA. Semiconductor clusters in the sol-gel process: Quantized aggregation, gelation, and crystal growth in concentrated ZnO colloids. Journal of the American Chemical Society. 1991; **113**:2826-2833

[34] Hjiri M, El Mir L, Leonardi SG, Pistone A, Mavilia L, Neri G. Al-doped ZnO for highly sensitive CO gas sensors. Sensors and Actuators B: Chemical. 2014;**196**:413-420

[35] Li H, Wei Y, Zhang Y, Zhang C, Wang G, Zhao Y, et al. In situ sol-gel synthesis of ultrafine ZnO nanocrystals anchored on graphene as anode material for lithium-ion batteries. Ceramics International. 2016:1-8

[36] Bøjesen ED, Jensen KMØ, Tyrsted C, Lock N, Christensen M, Iversen BB. In situ powder diffraction study of the hydrothermal synthesis of ZnO nanoparticles. Crystal Growth & Design. 2014;**14**: 2803-2810

[37] Demoisson F, Piolet R, Bernard F. Hydrothermal growth of ZnO nanostructures in supercritical domain: Effect of the metal salt concentration ($Zn(NO_3)_2$) in alkali medium (KOH). Journal of Supercritical Fluids. 2015;**97**: 268-274

[38] Wang R, Wu SP, Lv YC, Lin ZQ. Partially crystalline Zn_2GeO_4 nanorod/graphene composites as anode materials for high performance Lithium ion batteries. Langmuir. 2014;**30**(27): 8215-8220

[39] Bai S, Guo T, Zhao Y, Sun J, Li D, Chen A, et al. Sensing performance and mechanism of Fe-doped ZnO microflowers. Sensors and Actuators B: Chemical. 2014;**195**(3):657-666

[40] Alver Ü, Tanriverdi A. Boron doped ZnO embedded into reduced graphene oxide for electrochemical supercapacitors. Applied Surface Science. 2016;**378**:368-374

[41] Yoo R, Cho S, Song MJ, Lee W. Highly sensitive gas sensor based on Al-doped ZnO nanoparticles for detection of dimethyl methylphosphonate as a chemical warfare agent simulant. Sensors and Actuators B: Chemical. 2015;**221**:217-223

[42] Wang X, Huang L, Zhao Y, Zhang Y, Zhou G. Synthesis of mesoporous ZnO nanosheets via facile solvothermal method as the anode materials for Lithium-ion batteries. Nanoscale Research Letters. 2016; **11**(1):37

[43] Shen X, Mu D, Chen S, Wu B, Wu F. Enhanced electrochemical performance of ZnO-loaded/porous carbon composite as anode materials for lithium ion batteries. ACS Applied Materials & Interfaces. 2013;**5**(8): 3118-3125

[44] Azzaz Y, Kacimi S, Zaoui A, Bouhafs B. Electronic properties and stability of ZnO from computational study. Physica B: Condensed Matter. 2008;**403**(18): 3154-3158

[45] Pearton SJ, Norton DP, Ip K, Heo YW, Steiner T. Recent progress in processing and properties of ZnO. Progress in Materials Science. 2005; **50**(3):293-340

[46] Köse H, Dombaycıoğlu Ş, Aydın AO, Akbulut H. Production and characterization of free-standing ZnO/SnO_2/MWCNT ternary nanocomposite Li-ion battery anode. International Journal of Hydrogen Energy. 2016:1-9

[47] Ma D, Cao Z, Hu A. Si-based anode materials for li-ion batteries: A mini review. Nano-Micro Letters. 2014;**6**(4): 347-358

[48] Liu Y, Li Y, Zhong M, Hu Y, Hu P, Zhu M, et al. A facile synthesis of core-shell structured ZnO@C nanosphere and their high performance for lithium ion battery anode. Materials Letters. 2016;**171**:244-247

[49] Zhang L, Zhang J, Liu Y, Zheng P, Yuan X, Guo S. Al doped-ZnO nanoparticles implanted in reduced graphene oxide with improved electrochemical properties for lithium ion batteries. Materials Letters. 2016; **165**:165-168

[50] Dai J, Wang M, Song M, Li P, Zhang C, Xie A, et al. A novel synthesis of ZnO/N-doped reduced graphene oxide composite as superior anode material for lithium-ion batteries. SMM. 2016; **112**:67-70

[51] Brodie BC. On the atomic weight of graphite. Philosophical Transactions. Royal Society of London. 1859;**149**(9): 249-259

[52] Hummers W, Offeman R. Preparation of grafitic oxide. Journal of the American Chemical Society. 1958; **80**(6):1339

[53] Novoselov KS, Geim AK, Morozov SV, Jiang D, Zhang Y, Dubonos SV, et al. Electric field effect in atomically thin carbon films. Science. 2004;**306**(5696): 666-669

[54] Sattler KD. Carbon Nanomaterials Sourcebook: Graphene, Fullerenes, Nanotubes, and Nanodiamonds. Vol. I, 1st ed. CRC Press; 2016

[55] Hirata M, Gotou T, Horiuchi S, Fujiwara M, Ohba M. Thin-film particles of graphite oxide. 1: High-yield synthesis and flexibility of the particles. Carbon N. Y. 2004;**42**(14): 2929-2937

[56] Ehrentraut D, Sato H, Kagamitani Y, Sato H, Yoshikawa A, Fukuda T. Solvothermal growth of ZnO. Progress in Crystal Growth and Characterization of Materials. 2006;**52**(4):280-335

[57] Wang L, Tang K, Zhang M, Xu J. Facile synthesis of Mn-doped ZnO porous nanosheets as anode materials for lithium ion batteries with a better cycle durability. Nanoscale Research Letters. 2015:3-7

[58] Ur R, Sagar R, Mahmood N, Stadler FJ, Anwar T, Navale ST, et al. High capacity retention anode material for lithium ion battery. Electrochimica Acta. 2016;**211**:156-163

[59] Lee DW, De Los Santos VL, Seo JW, Felix LL, Bustamante AD, Cole JM, et al. The structure of graphite oxide: Investigation of its surface chemical groups. The Journal of Physical Chemistry. B. 2010;**114**(17):5723-5728

[60] Park RS, Ruoff S. Chemical methods for the production of graphenes. Nature Nanotechnology. 2009;**4**:217-224

[61] Lu XY, Jin XH, Sun J. Advances of graphene application in electrode materials for lithium ion batteries. Science China Technological Sciences. 2015;**58**(11):1829-1840

[62] Petnikota S, Rotte NK, Srikanth VVSS, Kota BSR, Reddy MV, Loh KP, et al. Electrochemical studies of few-layered graphene as an anode material for Li ion batteries. Journal of Solid State Electrochemistry. 2014;**18**(4):941-949

[63] Shan H, Li X, Cui Y, Xiong D, Yan B, Li D, et al. Sulfur/nitrogen dual-doped porous graphene aerogels enhancing anode performance of lithium ion batteries. Electrochimica Acta. 2016;**205**:187-197

[64] Vargas COA, Caballero Á, Morales J. Can the performance of graphene nanosheets for lithium storage in Li-ion batteries be predicted? Nanoscale. 2012; **4**(6):2083-2092

[65] Wan L, Ren Z, Wang H, Wang G, Tong X, Gao S, et al. Graphene nanosheets based on controlled exfoliation process for enhanced lithium storage in lithium-ion battery. Diamond and Related Materials. 2011;**20**(5–6): 756-761

[66] Miao C, Zheng C, Liang O, Xie Y. Chemical vapor deposition of graphene. In: Physics and Applications of Graphene—Experiments. 2011. pp. 37-54

[67] Kim KS, Zhao Y, Jang H, Lee SY, Kim JM, Kim JM, et al. Large-scale pattern growth of graphene films for stretchable transparent electrodes. Applied Physics Letters. 2009;**457**: 706-710

[68] Jang J, Son M, Chung S, Kim K, Cho C, Lee BH, et al. Low-temperature-grown continuous graphene films from benzene by chemical vapor deposition at ambient pressure. Scientific Reports. 2015:1-7

[69] Wu S, Xu R, Lu M, Ge R, Iocozzia J, Han C, et al. Graphene-containing nanomaterials for lithium-ion batteries. Advanced Energy Materials. 2015;**5**(21): 1-40

[70] Wu G, Jia Z, Cheng Y, Zhang H, Zhou X. Easy synthesis of multi-shelled ZnO hollow spheres and their conversion into hedgehog-like ZnO hollow spheres with superior rate performance for lithium ion batteries. Applied Surface Science. 2019;**464**: 472-478

[71] Yan J et al. Preparation and electrochemical performance of bramble-like ZnO array as anode materials for lithium-ion batteries. Journal of Nanoparticle Research. 2015; **17**(1)

[72] Pambudi YDS, Setiabudy R, Yuwono AH, Kartini E, Lee JK, Hudaya C. Effects of annealing temperature on the electrochemical characteristics of ZnO microrods as anode materials of lithium-ion battery using chemical bath deposition. Ionics (Kiel). 2019;**25**(2): 457-466

[73] Yang C et al. ZnO nanoparticles anchored on nitrogen and sulfur co-doped graphene sheets for lithium-ion batteries applications. International Journal of Ionics The Science and Technology of Ionic Motion. 2018;**24** (12):3781-3791

[74] Chowdari BVR, Srikanth VVSS, Adams S, Puttapati SK, Reddy MV, Gedela V. Unique reduced graphene oxide as efficient anode material in Li ion battery. Bulletin of Materials Science. 2018;**41**(2)

[75] Dong X, Yang Z, Yang W, Wu X, Lu Y. ZnO-templated N-doped holey graphene for efficient lithium ion storage performance. Materials Chemistry and Physics. 2018;**205**:487-493

[76] Wu J, Chen C, Hao Y, Wang C. Enhanced electrochemical performance of nanosheet ZnO/reduced graphene oxide composites as anode for lithium-ion batteries. Colloids and Surfaces A: Physicochemical and Engineering Aspects. 2015;**468**:17-21

[77] Fang H, Zhao L, Yue W, Wang Y, Jiang Y, Zhang Y. Facile and large-scale preparation of sandwich-structured graphene-metal oxide composites as anode materials for Li-ion batteries. Electrochimica Acta. 2015;**186**:397-403

[78] Guo R, Yue W, An Y, Ren Y, Yan X. Graphene-encapsulated porous carbon-ZnO composites as high-performance anode materials for Li-ion batteries. Electrochimica Acta. 2014;**135**:161-167

[79] Hsieh C-T, Lin C-Y, Chen Y-F, Lin J-S. Synthesis of ZnO@graphene composites as anode materials for lithium ion batteries. Electrochimica Acta. 2013;**111**:359-365

[80] Kim H, Jae W, Song J, Kim J. Skein-shaped ZnO/N-doped carbon microstructures as a high performance anode material for lithium-ion batteries. Journal of Alloys and Compounds. 2019; **772**:507–515

[81] Feng Y, Zhang Y, Song X, Wei Y, Battaglia VS. Facile hydrothermal fabrication of ZnO–graphene hybrid anode materials with excellent lithium storage properties. Sustain. Energy Fuels. 2017;**1**(4):767-779

[82] Qi Y, Zhang Y, Liu C, Zong S, Men Y. Room-temperature synthesis of ZnO @ GO nanocomposites as anode for lithium-ion batteries. Journal of Materials Research and Technology. 2018;**33**(10):1506-1514

[83] Jacob L, Prasanna K, Vengatesan MR, Santhoshkumar P, Woo C, Mittal V. Binary Cu/ZnO decorated graphene nanocomposites as an efficient anode for lithium ion batteries. Journal of Industrial and Engineering Chemistry. 2018;**59**:108-114

[84] Shen X, Cao Z, Zhang J, Li T, Jiang W. In-situ loading of ZnO nanoparticles on carbon felt as novel binder-free flexible anode for high performance lithium-ion batteries. Materials Letters. 2018;**229**:93-97

[85] Kim T, Kim H, Han JM, Kim J. ZnO-embedded N-doped porous carbon nanocomposite as a superior anode material for lithium-ion batteries. Electrochimica Acta. 2017;**253**:190-199